Tamim Younos, a water scientist, is founder and president of the Green Water-Infrastructure Academy. He earned doctoral degree in urban and environmental engineering from the University of Tokyo. He is a former research professor of water resources at Virginia Tech and serves as an adjunct professor at the University of the District of Columbia. He has published more than 160 research and technical articles and edited ten books on water science and management topics. His major recognitions include; American Rainwater Catchment Systems Association (ARCSA) Lifetime Achievement Award, American Water Resources Association (AWRA) Icko Iben Award, Friend of UCOWR (Universities Council on Water Resources) Award, Fulbright Scholar (Specialist) Award, and Japan Society for Promotion of Science Fellow Award.

To Yumiko and my family and friends who have encouraged and inspired my work throughout the years!

Tamim Younos

WATER AND ENERGY KNOWLEDGE FOR CITIZEN EDUCATION

Collection of Essays

AUSTIN MACAULEY PUBLISHERS™
LONDON • CAMBRIDGE • NEW YORK • SHARJAH

Ordering Information
Quantity sales: Special discounts are available on quantity purchases by corporations, associations, and others. For details, contact the publisher at the address below.

Publisher's Cataloging-in-Publication data
Younos, Tamim
Water and Energy Knowledge for Citizen Education

ISBN 9798891551664 (Paperback)
ISBN 9798891551671 (ePub e-book)

Library of Congress Control Number: 2023922424

www.austinmacauley.com/us

First Published 2024
Austin Macauley Publishers LLC
40 Wall Street, 33rd Floor, Suite 3302
New York, NY 10005
USA

mail-usa@austinmacauley.com
+1 (646) 5125767

The contents of this book are mostly extracted from the author's published technical reports and research articles, modified as short essays for easy reading useful for citizen education purposes. Most articles were published as opinion-educational columns on water science and environment in the Roanoke Times over a period of 10 years. Special acknowledgements are due to Tammy E. Parece, who assisted with editing many columns, and family and friends "fun club" who encouraged and inspired the author to write opinion pieces on water science and environment.

Table of Contents

Foreword

During the third decade of the 21st century, it's an appropriate time to make some new resolutions. What problems we want and need to address, and what do we want to achieve and accomplish? To move forward, first we need to look back and remind ourselves why we are here and how we got here! Our thinking and values have evolved over time – decades and sometimes generations. The 18th century Industrial Revolution resulted in the emergence of high-population urban centers and an agricultural sector which demanded more water and energy. As a result, uncontrolled volumes of contaminated domestic, industrial and agricultural wastes were discharged into the environment which caused significant water, air and soil pollution. The environmental values of the 18th century were dictated by the 18th century state-of-knowledge, and did not foresee unintended consequences on human health and ecosystem degradation which became more critical around the middle of the 20th century. Evolving environmental values are mostly based on advances in science and technology.

The environmental revolution in America was triggered in 1960s. For example, Rachel Carson's book, *Silent Spring*, published in 1962, brought significant public awareness about DDT and other pesticides' impact on human health and environment. Public awareness and activism led to enacting environmental laws and regulations to protect waters of the United States. The U.S. Environmental Protection Agency (EPA) was created in 1970; the Clean Water Act of 1972 expanded the 1948 Federal Act to regulate pollutants' discharge into surface waters and establish water quality standards for surface waters; the Safe Drinking Water Act of 1974 was established to ensure the safety of public drinking water supplies. Before the 1960s, the focus of water research and education in America was mostly on water source development and delivery. In 1970s, the focus shifted to non-point source pollution, i.e., stormwater runoff from agriculture and urban areas.

In the twenty-first century, urban environments are facing serious challenges inherited from the past centuries. These include but are not limited to: (1) anthropogenic water scarcity due to convergence of population in urban areas; (2) competition for water demand for food production and industrial use; (3) energy dependency of urban water infrastructure; (4) aging water infrastructure; (5) consequences of a changing climate affecting drinking water quality and availability; (6) emerging contaminants (e.g., pharmaceuticals, hormones, PFAS) in natural water systems and drinking water; and (7) drinking water cost and equitable water utility management.

Water problems identified in the 20th century still continue to impact the environment to a great extent. And new problems emerged during the first two decades of 21st century. In the aftermath of 9/11 terrorist attack in 2001, the beginning of the first decade of the 21st century, cybersecurity in water infrastructure began to dominate water research, education and management. During the second decade of the 21st century, the impact of climate change on water resources and water infrastructure management and policy became a prominent issue and continues to this day. Most recently, COVID-19, the unexpected crisis, has exposed a diverse range of preexisting water infrastructure problems, particularly poor drinking water quality in low-income communities.

From a water management perspective, we still utilize and mostly depend on traditional water management technologies, policies, and college curricula which were developed in the 20th century. And in recent decades, the strong interconnectedness of the global economy has significantly increased the importance of global climate variability and human activities as critical factors in sustainable management of water resources. Furthermore, understanding the social and human dimension of water management is emerging as the most significant imperative in equitable water management.

At present, there is a significant need to communicate scientific and objective information related to water issues to citizens and other stakeholders. Contents of this book are mostly extracted from the author's published technical reports and research articles and modified as short essays for easy reading useful for citizen education purposes. The book provides objectives information on science and technology of water management and energy use in the water sector, and innovative strategies for water and environment preservation to interested citizens, high school level students, science teachers, and other stakeholders.

1. Water Is Life

1.1 Water and Our Quality of Life in the 21st Century

Early human communities were established near a water source. Ancient civilizations used various types of vessels, made from animal skin or clay, to carry water from its source to royal palaces, peasant households and war zones. Science and technology of water resources development evolved over centuries. Using the gravity flow concept to extract surface and groundwater, and to transport to distant communities, dates back to 800 B.C.E. Persia. Archimedes screw, which used human/animal power to lift up water from ditches to higher agricultural fields, came into existence about 200 B.C.E. The steam powered pump was developed in the 19th century when coal became available as an energy source. Pumps replaced human/animal power to move water and facilitated efficient water source development and transportation to distant villages and towns. In some way, pumps are the original robots and the beginning of mechanization replacing manpower. Methods for measurement of properties of water flourished in the 19th century. The connection between water quality and public health, i.e., waterborne diseases such as cholera and dysentery, were also established in 19th century, thus the need and concept for water treatment before human consumption.

In the 20th century, urban population growth and land development changed the natural landscape and natural water flow characteristics. According to the USDA Forest Service, 'every day, America loses more than 4,000 acres of open space to development; that's more than 3 acres per minute.' Furthermore, increased water demand necessitated development of various water infrastructures: dams and reservoirs; exploitation of deep groundwater; centralized drinking water treatment and delivery network; sewer drainage, treatment and disposal network; and stormwater drainage network, to dispose runoff water from cities' paved surfaces and rooftops. Additionally, in the 20th century, increased water use in agriculture and fossil-fuel extraction and electricity generation caused significant pressure on water resources demand and development. There were unintended consequences of accelerated urbanization, industrialization and increased agricultural and energy production.

The impact of agricultural chemicals on water quality and human health were recognized in 1960s when the book "Silent Spring" was published (Rachel Carson, 1962). The environmental revolution was launched and led to significant regulatory steps to protect the U.S. water, soil and air environments. The U.S. Environmental Protection Agency (EPA) was created in 1970. The Clean Water Act of 1972 expanded the 1948 Federal Act to regulate pollutants' discharge into the U.S. waters and establish quality standards for surface waters. The Safe Drinking Water Act of 1974 regulates the safety of public drinking water supplies. The original Clean Air Act was enacted in 1963 with further amendments in 1970 (and 1990). No doubt, these regulations, based on the science and technology of the 20th century, have significantly contributed to improved environment and quality of life. But are regulations and traditional technologies developed in the 20th century adequate to meet significant challenges we're facing in the 21st century?

Major challenges of the 21st century include availability of adequate and safe drinking water free of contaminants such as lead, microbes, hormones and pharmaceuticals; enhancing deteriorating water delivery network; groundwater preservation; energy use efficiency; climate change and its consequences on water resources (droughts and floods) and water infrastructure safety; and increased competition for water demand in urban, agriculture and energy sectors.

There is significant need for a paradigm shift to meet these challenges for a better quality of life in the 21st century. This can be accomplished by implementing holistic and decentralized green water-infrastructure which considers the interconnectedness of natural and engineered systems in planning urban water-infrastructure. Green water-infrastructures are decentralized water systems that integrate locally available water and renewable energy resources for small-scale water treatment, water delivery, and use at the local level. Locally available water resources include rainwater, stormwater, groundwater, wastewater, salt and brackish water. Locally available renewable energy resources include solar, wind, micro-hydro, geo-thermal, biomass and other. These futuristic approaches are already underway in some locations in the U.S. and other countries. *"The future is already here. It's unevenly distributed"*. (William Gibson, Science Fiction Writer). Our major challenges to implement broader decentralized water-infrastructure include upgrading traditional college curricula in water management and water-infrastructure, informing local governments about possibilities and advantages, and promoting policy incentives and legislation for implementing decentralized green water-infrastructure.

1.2 The Intersection of Water Science and Society

The dictionary definition of intersection is "a place or area where two or more things cross (such as streets)". In term of water sciences, this is a relatively new concept. From a water management perspective, it means how technology with good intentions could affect the water systems with possible negative impacts. A good example is urban development where agricultural and forested areas are replaced with paved areas and buildings, thus affecting hydrology (natural water movement) of the area, i.e., surface water flow, evaporation, and underground infiltration of rainwater. Consequently, contaminated stormwater runoff collected from paved areas is discharged to streams, rivers and lakes and affecting the surface water quality and ecosystem. The term *intersection* includes understanding the impact of human activity on water availability, water quality and related fields such as land use management. The emerging scientific field is called social hydrology.

It is interesting and informative to note the progression of water science and technology overtime, which is as follows: *water source use – water source development – water source management – sustainable water management – resilient water management.* Original human communities were established in close proximity to natural water systems, i.e., streams and rivers, lakes, springs and oceans, where water was readily available for human use. Ancient civilizations developed water sources by means of gravity flow, using differences in elevation. For example, Persians, in the early part of the first millennium BCE, constructed elaborate tunnel systems called *qanats* for transporting water from mountainous areas to lower elevation villages for domestic use and irrigating field crops. Later, ancient technologies based on gravity flow progressed to water management. A good example is Istanbul's aqueduct and Basilica Cistern built during the 6th century CE. Water transported by aqueduct was stored in large cistern for eventual use.

The 18th century Industrial Revolution triggered significant human intervention in natural systems. Energy production and use, in the form of fossil fuels (i.e., coal and petroleum), enabled water transport from source to distant consumers, caused the emergence of high-density population (urban) centers, and accelerated agricultural and industrial production. The 20th century witnessed significant modernization in the arena of water resources development and water infrastructure – building dams and reservoirs; using powered pumps to extract deep groundwater and surface water resources; centralized water treatment plants; water delivery pumps and pipelines to transport clean drinking water to homes, and public and commercial buildings; and installing sewer and stormwater pipes to move wastewater and stormwater runoff away from population centers. The term "sustainable management of water resources", which included economic sustainability of water management was popularized in recent decades.

During the third decade of the 21st century, human societies across the world are being challenged with significant water-related problems. These problems are exacerbated by anthropogenic (human impact) climate change, a phenomenon that is largely accelerated due to human intervention in natural systems since the Industrial Revolution. The impact of climate change on water resources includes, but is not limited to, changes in precipitation patterns and intensity; the severity and length of droughts; sea level rise and associated consequences (e.g., flooding of coastal cities and encroachment of saline waters into freshwater aquifers). Obviously, technology and water management strategies implemented in the 20th century did not fully integrate social and anthropogenic factors in the planning and design of water management systems.

There is a significant need to shift toward resilient water management strategies. A major component of resilient water management strategy is increased public participation, interaction and feedback in water management and decision-making. The major challenge we're facing today is how to incorporate the concept of the intersection of water science and society into water policy and regulation, which are outdated and way behind advances in water science and technology. From a practical perspective, we can move forward with developing resilient water management strategies at the local level, for example, using locally available alternative water sources such as rainwater in urban agriculture, and using locally available renewable energy sources for water treatment and distribution at the local level, i.e., developing a decentralized green water-infrastructure strategy with increased societal awareness, public interaction and participation. There is also a significant need for upgrading college curricula on cross-disciplinary and resilient water management strategies.

1.3 U.S. Drinking Water Problems: Flint, Michigan and Beyond

Lead in Flint, Michigan drinking water was in the national news. However, a closer look at drinking water systems shows that the lead issue is just the tip of iceberg. At present, more than 80 percent of the U.S. population reside in urbanized areas and are served by more than 60,000 municipal drinking water treatment plants. The American Society of Civil Engineers (ASCE) 2013 Report Card noted that much of U.S. drinking water infrastructure is nearing the end of its useful life, and America's drinking water systems face an annual shortfall of at least 11 billion dollars to replace aging facilities. There are an estimated 240,000 water main breaks annually in the U.S., and approximately 23 million cubic-meters of potable water is lost due to leakage when treated potable water is transported via distribution pipelines from a water treatment plant to consumers. Furthermore, a 2001 study estimated that in the

U.S. nearly 7,000 water systems affecting about 10.5 million people violated microbial drinking water standards. The longer the water distribution pipeline, the higher is potential for microbial contamination of tap water.

Other major challenges facing drinking water systems include water source (surface and groundwater) degradation, emerging contaminants in source water and drinking water, high energy consumption (it is estimated that potable water and wastewater sector use about 75 billion kWh of electricity per year or about 4% of total energy use in the U.S.), and uncertainty and threat posed by climate change.

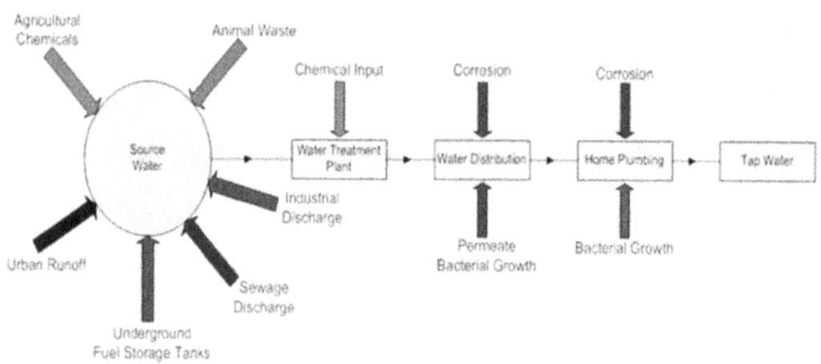

Source: Author

This figure shows potential sources of contaminants present in drinking water source. Conventional public water treatment plants are designed to remove contaminants from drinking water. The treated water is tested for over 90 contaminants (microorganisms, disinfectants, disinfection byproducts, inorganic Chemicals, organic Chemicals, radionuclides) before leaving the water treatment plant for transmission via pipelines to consumers in accordance to drinking water standards which is regulated by the U.S. Environmental Protection Agency.

Health problems associated with lead in drinking water originating in lead pipes are well documented. The U.S. Congress enacted the Reduction of Lead in Drinking Water Act on January 4, 2011, to amend Section 1417 of the Safe Drinking Water Act (SDWA) regarding the use of lead pipes, plumbing fittings or fixtures, solder and flux. The Act established a prospective effective date of January 4, 2014, which provided a three year timeframe for affected parties to transition to the new requirements. However, pockets of lead contamination of drinking water still exist across the U.S., and are best highlighted in Flint, Michigan.

At present, water utilities are facing many challenges regarding contaminants in drinking water and not limited to lead. Emerging contaminants in source water and drinking water are a critical health issue: emerging pathogens which are newly discovered microorganisms and/or newly recognized pathogens in water with less data available about their transmission routes, virulence, minimum infective dose and disinfectant susceptibility; endocrine disruptors which are chemicals that can interfere with the endocrine or hormone system in mammals and can cause cancerous tumors, birth defects, and other developmental disorders; pharmaceuticals which are undetected and find their way to drinking water; byproducts of chemicals used in water treatment, and byproducts of reaction between chemicals in treater water and pipe material.

Water treatment technology is an evolving science. Research and development of innovative and cost-effective water treatment technologies that address traditional and emerging contaminants in drinking water are underway. However, a major challenge is the design of contaminant free water transportation pipelines and replacement of old pipes which are cost prohibitive. A futuristic approach will include design and implementation of decentralized drinking water systems that integrate recovery of locally available wasted waters (stormwater and wastewater) and using small-scale advanced community-based water treatment technologies. Decentralized systems will reduce water transport pipeline length, reduce potential for drinking water contamination via pipes, use less electricity for transporting water, and consequently result in lower drinking water infrastructure cost to consumers.

1.4 U.S. Drinking Water Problems: Flint, Michigan and Beyond: Part II

In the previous section, I mentioned the existence of pockets of communities across the U.S. where there could be problems with availability of safe drinking water. To understand the problem better, it's important to know how the U.S. population obtains their drinking water. At present, about 50 million people in the U.S. depend on private water sources, mostly water wells for domestic needs. These private water supplies are not regulated. Homeowners are responsible for availability and safety of their drinking water, which may mean the need to dig the well deeper, install water treatment units, and/or get their water regularly tested. The real extent of water quality problems in private water supplies is unknown. However, significant water quality problems are reported where private water well supplies are tested.

According to the U.S. Safe Drinking Water Act, a "public water system" is an entity that provides 'water for human consumption through pipes or other constructed conveyances to at least 15 service connections or serves an average of at least 25 people for at least 60 days a year.' Public refers to people served not the system ownership. A public water system (PWS) may be government or privately owned. At present, 151,000 public water systems serve about 90% of U.S. population. PWS are classified according to the number of people they serve, the source of water, and whether they serve the same customers year-round or on an occasional basis. The population served by PWS range from several hundred thousand in large metropolitan areas to very small systems serving only 25 people or 15 connections. The great majority (97%) of PWS are small drinking water systems serving 10,000 or fewer people. PWS are regulated by the USEPA. To ensure PWS meet drinking water standards, the treated water should be tested for over 90 water quality parameters at the point where water exits the treatment plant. In Virginia, drinking water regulations are enforced by the Virginia Department of Health.

Public water systems, large and small, are facing many challenges. Deteriorating water infrastructure, particularly pipeline networks, is a significant problem. Numerous pipe breakdowns and corrosion significantly impact water availability and water quality across the U.S. on a daily basis. According to the American Society of Civil Engineers, U.S. drinking water systems face an annual shortfall of at least $11 billion to replace aging facilities that are near the end of their useful life. Emerging contaminants such as hormones and pharmaceuticals in water sources are causing significant health concerns and a lack of confidence in tap water quality.

Operation and maintenance costs of water utilities are mostly financed by water rates charged to consumers. Therefore, drinking water problems are more significant in small PWS with less financial resources, and limited technical and managerial skills for operation and maintenance, and to keep up with the required monitoring of finished water.

In recent years, significant progress is being made in small-scale advanced water treatment technologies, which allows private water supplies and community-based small PWS to attain safe drinking water. Small-scale water treatment technologies, i.e., decentralized water supply organized around cost sharing small-scale water treatment cooperatives, can reduce the need for extensive pipe networks to transport water to consumers and will facilitate cost effective maintenance. Local sources of water may include captured rainwater, groundwater, recycled wastewater, and saltwater. Decentralized systems also reduce energy use for water delivery. It is estimated that the centralized water infrastructures in the U.S. use about 75 billion kWh of electricity per year, about 4% of total U.S. energy use. A great portion of energy use is attributed to long distance water delivery. Renewable energy use in water sector is an evolving technology and has been implemented in parts of the U.S. and other countries. Utilities should consider integrating renewable energy when upgrading their water treatment system. It's envisioned that, in the future, locally available renewable energy resources, such as solar and wind energy will be integrated in decentralized water treatment and supply of locally available water sources.

1.5 Emerging Contaminants in Our Waters

In recent decades, significant scientific progress in the arena of water quality analysis has enabled scientists to (1) detect new contaminants in our waters which may have actually existed over a long period of time; and (2) measure contaminant concentration (amount) in water at trace (very low) levels.

Emerging contaminants include but not limited to pathogenic (disease causing) microorganisms, estrogens/hormones, pharmaceuticals and household products which as yet little data are available about their impact on human health and ecosystem. The USEPA collects data and periodically updates the drinking water Contaminant Candidate List (CCL) (https://www.epa.gov/ccl) and it was last updated in 2016. Several candidate microbial pathogens and chemicals are listed.

Pathogenic microorganisms (bacteria, viruses, and protozoa) in natural waters – some are natural habitants – mostly originate from animal waste, sewer and septic systems (human waste) and urban stormwater runoff. Pathogens pose a risk to human health through various uses of water. Conventional water treatment technologies for disinfection were generally considered effective for removing pathogens from water, but in 1993, *Cryptosporidium parvum* (a disinfectant-resistant protozoan pathogen) in drinking water was the cause of the largest waterborne disease outbreak in U.S. history. This outbreak affected more than 400,000 people in Milwaukee, Wisconsin, and caused more than 100 deaths. Waterborne disease outbreaks associated with recreational water are common and attributed to the use of public pools, hot tubs, rivers, lakes, beaches, and water fountains. Shellfish contaminated via waterborne routes can impact human health through the food chain such as oyster consumption.

In 1999–2000, the US Geological Survey (USGS) conducted the first national-scale examination of emerging chemical contaminants in U.S. waters. The study documented the presence of low levels of many emerging contaminants, including prescription and non-prescription drugs, hormones, and other contaminants, in a network of 139 targeted streams across the United States. As a follow up, the USGS recently conducted two national-scale reconnaissance studies to collect baseline information on the environmental occurrence of pharmaceuticals, personal-care products, detergents and flame retardants at 25 groundwater and 49 surface water sites in 25 states and Puerto Rico. According to study report, pharmaceuticals, including antibiotics and prescription and non-prescription drugs, generally were detected less frequently in sources of drinking water than they were in the national stream reconnaissance study of 1999–2000. The USGS study concluded that 'it is important to note that detection at a low concentration does not necessarily signal a health concern, and that some of the chemicals detected in these reconnaissance studies can occur naturally.'

The USEPA regulations, under the Safe Drinking Water Act, require utilities to identify the best available technology for treatment of contaminated source waters. Drinking water is tested at the exit of water treatment plant to ensure compliance with Drinking Water Standards. However, less is known about emergence of microbial and chemical pollutants in water distribution pipelines. The challenge is how to balance the need for adequate disinfection (chlorination) while reducing the potential chronic health effects of chlorine and its byproducts. Recently, there is a trend to use chloramines instead of free chlorine as disinfectant, since there is less potential for generating harmful byproducts. However, little is known about potential health effects of switching to chloramines that could arise from increased microbial survival or growth in the drinking water distribution system. Furthermore, research has indicated that a switch from free chlorine to chloramines disinfectant triggers lead release from home plumbing pipes, which pose significant health risk to society.

Alarm bells were rung in 2004 when traces of the antidepressant Prozac were detected in London's drinking water. Legal and illegal drugs/opioids disposed into the sewer system via human urine, are not detected or removed at wastewater treatment plants and are discharged to sources of drinking water – rivers/streams and lakes. These emerging contaminants are not detected at conventional drinking water treatment plants and may ultimately find their way in tap water. The health impact of low level second-hand drug/opioids consumption by general population is unknown. How scientific research and advances in water treatment technologies are effectively integrated into holistic water resource management to protect societal health and ecosystem remains a significant challenge in the 21st century.

1.6 The Saga of COVID-19

In our complicated world demonstrated by global spread of COVID-19 (coronavirus), one eternal fact "Water is Life" still remains the most critical fact to fight the deadly virus. According to the Centers for Disease Control and Prevention (CDC), 'The COVID-19 virus has not been detected in drinking water. Conventional water treatment methods that use filtration and disinfection, such as those in most municipal drinking water systems, should remove or inactivate the virus that causes COVID-19.' Wash your hands with clean water and soap is considered a life-saving action for self-protection and controlling the germ spread. We, in the U.S., are fortunate – a significant majority of people have access to clean water to wash our hands. According to the World Health Organization (WHO), 1 in 3 people globally do not have access to safe drinking water. Some 2.2 billion people around the world do not have safely managed drinking water services – drinking water from sources that are free from contamination and available when needed. And about 3 billion people lack basic handwashing facilities – having a protected drinking water source that takes less than thirty minutes to collect water from and having handwashing facilities with soap and water in the home.

The corona virus pandemic will pass but it's likely that the U.S. and the world will face other pandemics in the future. The million-dollar question is: are we really prepared? Current U.S. population – over 330 million – demands significant volumes of water for domestic, agricultural and industrial consumption. But we also affect the quality of available freshwater resources, as well as ocean waters, due to discharge of low quality used-water to these water bodies. In 2019, the American Water Works Association (AWWA) published the State of the Water Industry (SOTWI) survey. The report lists 30 major problems that water professionals in the U.S. face in providing safe drinking water. Most critical challenges noted in the report are (1) long-term water supply availability; (2) watershed and source water protection; (3) groundwater overuse and management; (4) water conservation; (5) drought or periodic water shortages; (6) water loss control; (7) energy use efficiency and cost; (8) expanding water reuse and reclamation; (9) climate risk and resiliency; and (10) cybersecurity of large water infrastructure. On the top of the list are problems associated with aging water infrastructure and financing of the water infrastructure.

Water infrastructure problems, noted above, demand a paradigm shift toward innovative approaches to effectively cope with existing and emerging problems. For example, a major component of the futuristic water management will be integrating the reuse of locally available alternative water sources into water management systems. According to the EPA Water Reuse Action Plan (2019), possible sources of water for potential reuse include municipal wastewater, industry process and cooling water, stormwater, agriculture runoff and return flows, and oil and gas produced water. Examples of reuse applications include agriculture and irrigation, potable water supplies, groundwater storage and recharge, industrial processes, onsite non-potable use, saltwater intrusion barriers, and environmental restoration. Rooftop rainwater capture and use, which is not included in the EPA water reuse category, also has significant potential as an alternative water source in urban areas – building rooftops constitute 30–40% of impervious areas in urban settings – for various uses including indoor potable and non-potable uses. Major advantages of using alternative water sources are (1) reducing demand on natural freshwater sources and less discharge of low-quality water into natural water bodies; and (2) less pressurized pipe network and consequently less energy use.

Coronavirus crises bring significant opportunity to reevaluate our direction and priorities in water management. As we celebrate the 50th anniversary of Earth Day on April 22, we need to pay special attention to water, our planet's most precious resource. Immediate tasks include upgrading college curricula that include innovative concepts for water management, promoting water policy incentives and regulations for holistic water management at the local, state and national levels, and citizen education and activism to be directed toward water conservation and meeting futuristic goals in water resources management.

1.7 Coping with Water Poverty in the Era of COVID-19

The COVID-19 crisis has exposed a diverse range of preexisting community problems across the U.S. which has significantly added to the calamity of COVID-19. Typical problems include various levels of crisis management deficiency at the local level, such as lack of safe drinking water supplies. In two commentaries titled, "U.S. Drinking Water Problems: Flint, Michigan and Beyond", (Sections 1.3 and 1.4 above), the existence of pockets of communities across the U.S. where there could be problems with availability of safe drinking water were discussed. Amid the COVID-19 crisis, the water poverty problems are further exposed and covered in the national news.

In U.S., about 86% of population is served by public water systems (PWS). The remaining 14% of population depends on private water supplies (mostly groundwater wells). There are approximately 155,000 PWS in the United States. The U.S. Safe Drinking Water Act (SDWA) define "public water system" as an entity that provides 'water for human consumption through pipes or other constructed conveyances to at least 15 service connections or serves an average of at least 25 people for at least 60 days a year.' Public refers to people served not the ownership. A public water system may be publicly or privately owned. The serving population for PWS range from several hundred thousand in large metropolitan areas to very small systems that may serve only 25 people or 15 connections. The great majority of PWS are small drinking water systems that serve 10,000 or less people. PWS are regulated by the EPA in partnership with states and governed through a local government agency or a water authority. Water rates are designed to cover the water utility operation and water infrastructure improvement costs.

Small PWS across the U.S. are facing significant challenges. Small PWS usually have fewer financial resources (most often a low tax base) and needed technical and managerial skills for operation and maintenance of water systems, and, therefore, are unable to meet the EPA required monitoring of finished water and are susceptible to contamination. In 1996, the SDWA was amended to protect Americans from unsafe drinking water and to prevent contamination of drinking water supplies. Specifically, section 1420 of the SDWA focuses on developing the financial, managerial, and technical capacities of small water systems where violations of drinking water standards are prevalent. Many small public water systems could benefit from revenue growth or cost cutting ventures available to larger water systems, but small systems do not have the access to capital or are limited by geographic location.

In the past, our two separate studies at Virginia Tech investigated various options for consolidation and restructuring of small water systems and proposed a strategy for capacity development and restructuring of public small water systems in Virginia. After careful consideration of the different approaches by public and private organizations, the needs and motivations of those interested in capacity development, and the available resources, we concluded that a new approach, i.e., the implementation of a co-operative operation and management for small water systems will be most suitable in Virginia. The co-operative as a body will decide and implement the most cost-effective operational, administrative, managerial, and technical options for its members. Co-operatives have been used in many industries, mainly in agriculture, to protect and improve on utilizing scarce resources and to guarantee returns to all interested parties based on the level of use. Much in the form of the agricultural co-operatives, the water systems co-operative is an attempt to pool the resources of many small water systems (small businesses) to reach their objectives. First, there is a need to organize small water systems into a network of small businesses and educate the members on the benefits of co-operatives. The network can be used as a stepping stone toward the legal formation of a shared-service, non-profit small water system cooperatives. Drinking water supply problems in conjunction with the COVID-19 crisis provides a significant opportunity to restructure small drinking water systems. Safe drinking water is a basic human right.

1.8 Bottled Water: Panacea or Plague?

Early civilizations used various types of vessels, made from animal skin or clay, to carry water from its source for consumption in royal palaces, peasant households and war zones. At present, the synonym "bottled water" refers to containers of various sizes that provide drinking water to consumers. The modern water bottling industry was launched in 1960s when plastic was invented. However, the industry mostly flourished in 1990s when polyethylene terephthalate (PET) plastic became available. Packaging and transportation of bottled water became more feasible with PET plastic, because of its light weight and strength. Today bottled water is a major global commodity, a $22 billion industry that ranges from very small local bottling operations to giant international corporations.

Although the cost of typical bottled water per gallon is several hundred times higher than tap water, its popularity is mostly attributed to consumer preferences and perception. Many consumers perceive bottled water as "cleaner" and/or "healthier" than tap water. It's an acceptable substitute to sugary soft drinks and carried easily. In rural and isolated communities, bottled water availability is a necessity because of the absence of public water supplies and less confidence in quality of privately owned water wells and small water supply systems. Bottled water is an absolute necessity during emergency conditions when public water supplies are disrupted due to natural disasters or man-made events. Think Hurricane Katrina and Flint, Michigan.

Health and environmental concerns related to bottled water production and consumption include bottled water quality – for example, research has shown the adverse effects of bottled water storage duration and temperature on bottled water quality, plastic pollution caused by disposal of used plastic bottles, air pollution due to incineration of used plastic bottles, and significant energy used for plastic bottle production and bottled water transportation.

Bottled water production is regulated by the U.S. Food and Drug Administration (FDA) as a packaged commodity (food). Noted inadequacies of bottled water regulation are: 1) the FDA requires testing only once a year for bottled water quality, while in comparison, the USEPA mandates daily water quality testing and frequent monitoring of public drinking water supplies for contaminants; 2) the FDA does not have the specific statutory authority to require bottled water industry to use certified laboratories for water quality tests or to report test results, even if violations of the standards are found; and 3) the FDA's bottled water labeling requirements are similar to labeling requirements for other foods, but the information provided to consumers is less than what USEPA requires of public water supplies under the Safe Drinking Water Act.

To cope with plastic pollution, at present, recycling of used plastic bottles is considered the most appropriate management option compared to other practices (landfill disposal and incineration). However, recycling of plastic bottles is not yet a common practice. In the U.S., less than 30 percent of used PET bottles are recycled. Biodegradable plastic bottle has been introduced in the market and is expected to alleviate plastic pollution in the future.

Despite its high cost and noted environmental concerns, bottled water consumption in the U.S. shows increasing trend and nearly surpass soft drink consumption. A significant opportunity can be envisioned to enhance the societal role of bottled water. It can be promoted as a decentralized water production system for community development and job creation in low-income and inner urban areas as well as in affluent coastal cities, islands and resort areas where bottled water is transported from elsewhere. The decentralized system uses locally available water sources such as captured rainwater, grey water, and saltwater and integrates small-scale advanced water treatment technologies to produce high quality bottled water for local consumption. Local production of bottled water will reduce bottled water cost which is mostly attributed to energy used for transportation. Futuristic decentralized renewable energy technologies for water treatment will advance local bottled water production. Decentralized bottled water production could become a viable future solution to meet drinking water demand in urban as well as rural and isolated areas, create jobs at the local level, and enhance community development.

1.9 Revisiting the Spiritual Value of Water

On the planet Earth, there are many nationalities, numerous cultures and languages, people of different colors, different DNAs, and various levels of physical power and intelligence. Regardless of all those differences, all humans have one thing in common; an adult's body weight is about 50–60 % water depending on age and gender. The human brain is composed of 90% water. Equity is reflected in humankind creation or evolution, whichever we tend to believe. The human body needs water to function and can live only a few days without water. It should also be noted that animals contain on average 60% water and vegetables up to 75%. Therefore, water is a necessity in all aspects of our daily life. Water is life!

Water and history of humans are intertwined. Ancient civilizations were fascinated with water. For example, according to Goddess-Guide.com, *'The Water Goddess is found in all ancient cultures, reflecting the importance of this precious element in everyday survival. There are Goddesses that represent water in all its forms from sacred wells and lakes to the seas and immense oceans that cover our beautiful planet.'* The International Union for Conservation of Nature (IUCN), Environmental Law Centre (Léah Khayat and Diego Jara, April 2021) citing various references noted the following about the spiritual value of water: *'In Aztec mythology, the goddess Chalchiuhtlicue represented the entire range of the natural manifestations of water and as such she was involved in every part of life, from birth to death. For the Maya, cenotes (underground chamber or cave filled with water) were the gateway to another world, where supernatural beings live and where the souls of the dead go. This divine connection with water manifested itself also through the development of installations aimed at conserving and managing water for ceremonial, economic and social purposes, such as the crop terraces and irrigation systems in the Inca Sacred Valley or the Nazca aqueducts in Peru.'*

From a religious perspective, water is considered a gift of God and a spiritual symbol shared by the world's major religions. Overall, water is mentioned over 700 times in the Old and New Testaments. The Old Testament (Genesis 1:1), 'In the beginning… the Spirit of God was hovering over the waters.' In Islam, water is also considered a gift of God and symbol of purification. In Hinduism, it's believed that water purifies the body and the mind, and rejuvenates the spirit. Immersion in the Ganges River is thought to purify humans of their sins and, if a person's ashes are scattered into the river, it facilitates liberation from an eternal cycle of life and death, i.e., reincarnation. *'In Buddhism, water symbolizes life, the purest form of food, and water is the particular element which in nature carries everything together. Water symbolizes purity, clarity and calmness, and reminds us to cleanse our minds and attain the state of purity* (Trudy Fredriksson, The Swedish Buddhist Community).'

Modern science is agreement with world's religions regarding the origin of life and recognizes that life began in water. Scientists continue to search for water on other planets because water presence can be an indication of presence of some form of life, at least microbial life. As noted, in ancient civilizations, there was an obvious interconnectedness between science and spiritual value of water. But now in our modern world, unlike ancient civilizations, there is a significant disconnectedness – the spiritual value of water is ignored in water use and water source protection. Human intervention in natural systems since the Industrial Revolution has significantly harmed the water ecosystem. We do not respect water as the source of life and waste lots of water in various ways. We attempt to solve human generated water quality problems (for example, lead in drinking water) solely by developing new technologies or engineered systems which often generates new water quality problems. The enormous use of chemicals in water treatment systems is a good example. Will our generation, in the 21st century, be able to reconcile science and the spiritual value of water through integration of natural and engineered systems and other appropriate actions? This question should be subject to open debate and discussion.

2. Water Management Challenges and Futuristic Approaches

2.1 The Symphony of Water Management System

An orchestra symphony is composed of many pieces using several musical instruments – wind, string, brass and percussion – and aims to perform in harmony. The symphony of water management is composed of several engineered pieces that need to perform in harmony in order to preserve the health of humans and the environment. However, existing water management pieces are designed on separate tracks where each piece performs independently, and very often are in conflict with each other.

A system implies several components. Major components of a water management system are water source, water infrastructure, and water use. Water sources include fresh surface waters (streams, river and lakes), groundwater (wells and springs), rain water, and saline water. Water infrastructures are engineered systems that facilitate various uses of water resources, aim to protect human health and environment, and mitigate natural disasters such as floods and droughts. Water uses include potable water for domestic and commercial sectors, water for manufacturing and industrial sector including energy production, and agricultural use – crop and livestock/poultry production. Of course, the most critical use of natural waters relates to fisheries and recreational activities. Freshwater water resources are limited; its availability is affected not only by natural and geographic distribution but also by water demand. One of the critical challenges we face in the 21st century is increasingly competing demands for various uses of water.

Water use is the key component of water management system. Better planning of water use and use efficiency will lead to less conflict between various uses of water and the optimum design of water infrastructure. Questions to be considered include: 1) how a particular water use could negatively impact the potential for other uses of water; 2) how to prevent the impact of certain uses of water on water quantity/quality of available water sources; and 3) how to achieve the 21st century water quantity and quality goals in a cost-effective fashion. The answer to these questions lies in the concept of water reuse, and the concept of integrating engineered and natural system in the design water infrastructure.

On April 28, 2019, the USEPA issued the *"Discussion Framework for Development of a Draft Water Reuse Action Plan" and Vision for public review.* It states that *'Water reuse can be a valuable means to enhance the availability and effective use of our Nation's water resources and should be considered as part of an integrated water resources management approach to meet the future needs of our Nation. An integrated approach commonly involves a combination of water management strategies (e.g., water supply development, water storage, stormwater management, water use efficiency, and water reuse) and engages multiple stakeholders and needs… For purposes of this discussion framework, "water reuse" includes other common terminology including recycled water, reclaimed water, alternative water supplies, improved water reliability, and water resource recovery.'*

Several articles in this book discuss topics relevant to the proposed Framework (For example see, "Imperatives for 21st century water management and infrastructure" and "Water and our quality of life in the 21st century". Rainwater/stormwater is a major alternative water source and its appropriate use is one of the key solutions. Rainwater/stormwater "wasted water source" can be and should be used at the local level. Potential uses of rainwater/stormwater include but not limited to: 1) indoor and outdoor non-potable uses at schools, hotels and dormitories, churches; 2) urban community gardens; 3) aesthetic uses such as creating ponds; 3) water for livestock and poultry production; 4) and groundwater recharge, particularly in coastal aquifers to prevent salt water intrusion. With appropriate treatment, rainwater can also be used as potable water where needed.

A knowledge-based approach to rainwater/stormwater use will reduce the consumption of river and lake waters thus keep surface waters intact for fisheries and other uses, reduce surface water pollution originating from urban paved areas, preserve groundwater for coping with potential drought, and reduce flooding. Furthermore, decentralized water infrastructure at the local level using a local water source will reduce dependency of water use on energy, an effective mitigation strategy to cope with climate change. Composing the symphony of water management system in harmony with nature at the local level should be our major goal in the 21st century.

2.2 Imperatives for 21st Century Water Management and Infrastructure

The 20th century was the era of significant modernization in the arena of water resources development and water infrastructure. The past century accomplishments include building dams and reservoirs, developing groundwater resources, constructing centralized water treatment plants and water delivery infrastructure (pipelines and pumps) to make tap water available in homes and buildings, installing sewer and stormwater pipes to move away wastewater and stormwater from population centers and eventual discharge to surface waters, and using significant amount of fossil-fuels based energy to meet the critical need for water source development, water treatment and long-distance delivery to consumers. However, technologies and strategies that were developed based on the 20th century state of knowledge are no longer adequate and effective to meet emerging challenges of this century, particularly in urban environments where over 80 percent of U.S. population live.

The combined effects of urbanization/population concentration and consequent high water demand, documented in literature, include but are not limited to: (1) lowered groundwater table due to combined effects of excessive groundwater withdrawal and reduced natural groundwater recharge – low natural infiltration caused by increased paved areas; (2) saltwater intrusion in coastal aquifers impacting groundwater quality where nearly 50% of U.S. population live; sea level rise in coastal cities such as Norfolk, Virginia; and (3) widespread pollution of surface waters – rivers/tributaries and lakes caused by stormwater runoff. Significant problems in water infrastructure include but are not limited to overall deterioration of water infrastructure (breakdowns and leakage); lead and emerging contaminants such as antibiotics and hormones in drinking water; anthropogenic (man-made) conditions causing water scarcity and flood in urban areas; the need to cope and adopt to climate change and extreme weather conditions; reduce significant energy use and loss via water distribution systems; and cyber security concerns in large/centralized drinking water infrastructure.

Challenges noted above demand a paradigm shift towards innovative approaches to effectively cope with existing and emerging problems. A need exists for developing novel holistic approaches that integrates natural and engineered systems into planning and design of urban water-infrastructure system; and solutions that recognize the nexus between water, energy and food production in urban environments. Four major pillars of urban water management and water infrastructure in the 21st century are green technologies and low impact development – approaches that support water source conservation and ecosystem/groundwater preservation; urban aesthetics such as stream daylighting and artificial ponds; using alternative water resources that include rainwater/stormwater runoff, wastewater and brackish/saltwater; and shift toward integrating renewable energy resources into water infrastructure. The holistic approach can be best realized by incorporating small-scale decentralized water and energy production systems into urban environments. Decentralized concept is based on maximizing the use of locally available water and renewable energy resources. Furthermore, integrating decentralized urban food production systems into water and energy nexus can be an innovative solution that supports sustainable living and community development in urban areas.

Major impediments to holistic water management approach in urban environments include: (1) traditionally conservative attitudes within water resources engineering/planning community; (2) silo-mode college curricula in water management and infrastructure; (3) regulatory and policy gaps and hurdles; and (4) public perception. A simple illustrative example is the issue of urban stormwater management where stormwater drainage networks are continuously expanded with less consideration for alternative green technology approaches. To some extent, this impediment can be associated with silo-mode college curricula which fail to prepare future engineers and planners in integrated cross-disciplinary approaches in water management: engineering, hydrologic sciences, chemical and biological sciences, plant and food production sciences, geospatial technologies, information technologies, cyber infrastructure, and socio-economic sciences.

In general, policy making at state and local level does not keep up with technological advances. There is a significant need for updating policies and regulations, such as changes in zoning ordinances and building codes, and economic incentives to promote implementation of green technologies into retrofits and new developments. Last, but not least, public perception remains a serious impediment in urban water management. Thus, a focus on citizen education and K-12 curricula should be a top priority to change public perception.

2.3 Imperatives in Water Management: Today to 2030

We've just entered the third decade of the 21st century and it's an appropriate time to make our new decade resolutions. What problems we want and need to address, and what do we want to achieve and accomplish? To move forward, first we need to look back and remind ourselves why we are here and how we got here! Our thinking and values have evolved over time – decades and sometimes generations. The 18th century Industrial Revolution resulted in the emergence of high-population urban centers and an agricultural sector which demanded more water and energy. As a result, uncontrolled volumes of contaminated domestic, industrial and agricultural wastes were discharged into the environment which caused significant water, air and soil pollution. The environmental values of the 18th century were dictated by the 18th century state-of-knowledge, and did not foresee unintended consequences on human health and ecosystem degradation which continued to the middle of the 20th century. Evolving environmental values are mostly based on advances in science and technology.

The environmental revolution in America was triggered in 1960s. For example, Rachel Carson's book "Silent Spring", published in 1962, brought significant public awareness about DDT and other pesticides' impact on human health and environment. Public awareness and activism led to enacting environmental laws and regulations to protect waters of the United States. The U.S. Environmental Protection Agency (EPA) was created in 1970; the Clean Water Act of 1972 expanded the 1948 Federal Act to regulate pollutants' discharge into surface waters and establish water quality standards for surface waters; the Safe Drinking Water Act of 1974 was established to ensure the safety of public drinking water supplies. Before the 1960s, the focus of water research and education in America was mostly on water source development and delivery. In 1970s, the focus shifted to non-point source pollution, i.e., stormwater runoff from agriculture and urban areas. Water problems identified in the 20th century still continue to impact the environment to a great extent, and some problems such as contaminated urban stormwater runoff have been intensified due to significant urban growth.

In the aftermath of 9/11 terrorist attack in 2001, the beginning of the first decade of the 21st century, water security issues began to dominate water research, education and management. During the second decade of the 21st century, the impact of climate change on water resources/water infrastructure management and policy became a prominent issue and continues to this day. At present, the beginning of third decade of the 21st century, COVID-19, the unexpected crisis, has exposed a diverse range of preexisting water infrastructure problems across the U.S., including poor drinking water quality in low income communities and the need for environmental justice. In many ways, problems emerged during the first two decades of 21st century are added to the problems we inherited from the 20th century.

Today, in the third decade of the 21st century, major water management challenges include the availability of adequate and safe drinking water free of contaminants such as lead, microbes, hormones and pharmaceuticals; enhancing deteriorating water infrastructure; groundwater and ecosystem preservation; energy use efficiency in all water sectors; climate change and its consequences on water resources (droughts and floods); cybersecurity for water infrastructure; and coping with increased competition for water demand in urban, agriculture and energy sectors. Food, energy and water (FEW) nexus, or the interrelationship, is the emerging research topic with significant consequences.

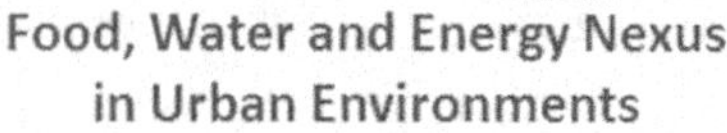

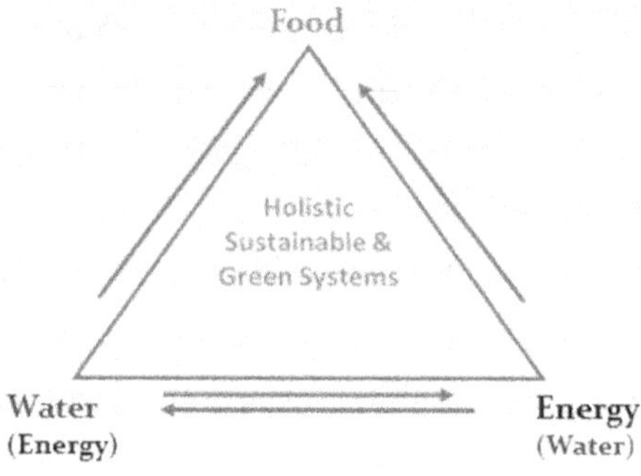

Source: Author

Unfortunately, from a water management perspective, we still mostly depend on traditional water management technologies, policies and college curricula which were developed in the 20th century and are, obviously, not adequate to meet significant challenges facing us today. In recent decades, the strong interconnectedness of the global economy has significantly increased the importance of global climate variability and human activities as critical factors in sustainable management of water resources. And there is a significant need for a paradigm shift in water research, education and regulations to meet water related environmental and societal goals. In the beginning of this decade, understanding the social and human dimension of water (and energy) management is emerging as the most significant imperative in water management.

2.4 Rethinking Water Management in Urban Areas

In early human civilizations, an understanding of gravity flow enabled humans to transport water from a river and shallow groundwater via man-made channels, qanats and aqueducts for domestic use and farming. Later, the Industrial Revolution and energy generation technologies enabled humans to defy gravity and use pumps to extract, lift up and move water via pressurized pipelines long distances and uphill, anytime and anywhere for a variety of uses.

In the 20th century, severe competition emerged, and still continues, over water demand for agriculture, industrial (including energy generation) and municipal uses of water. Water, food and energy security are high priorities for an acceptable standard of living. While water-use efficiency for food production has vastly improved due to technologies such as sprinkler irrigation, it remains energy intensive because it is highly dependent on electricity. However, in the 21st century, from a water availability perspective, the most critical global problem is high water consumption in urban areas. At present, 80.7 percent of total U.S. population, and 54 percent of the total global population reside in urban areas and show increasing trend. Urbanization and population growth necessitates the expansion of urban water-infrastructure – potable water supplies, wastewater treatment/discharge, and urban stormwater runoff control.

Urbanization and high water consumption have resulted in pollution of rivers and lakes, ecosystem degradation, low groundwater levels, and saltwater intrusion in coastal aquifers. Additionally, the modern urban water-infrastructures are highly energy dependent and consume 4–10% of the nation's total energy use, mostly attributed to water transport. It's estimated that about 20–30% of treated water is lost via water pipeline leakage during transportation resulting in significant water and energy loss. These infrastructures and related technologies, mostly developed in early 20th century, are very inefficient due to substantial wastage of water and energy resources and upgrading these infrastructures is costly. There is a significant need for a different approach or paradigm shift for urban water management. A holistic approach for water management provides a new thinking in the arena of planning urban water-infrastructures. The holistic approach recognizes the interconnectedness and oneness of water resources and engineered water-infrastructures: potable water supply, wastewater generation/treatment, and stormwater runoff control. These infrastructures are traditionally planned and designed as separate systems.

Integrated Decentralized Water and Renewable Energy Systems

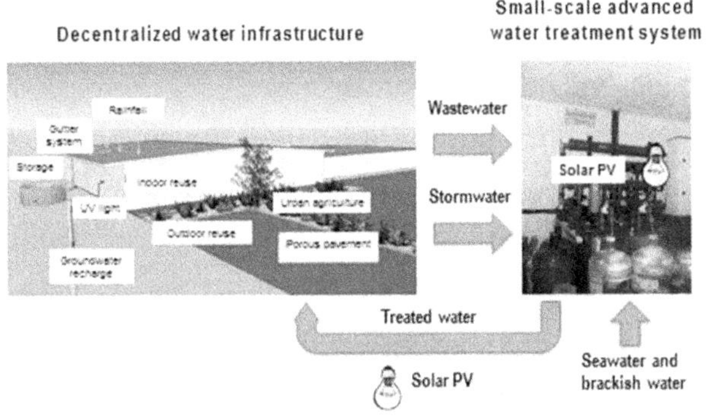

Source: Author

The holistic approach can be realized by incorporating decentralized water infrastructure concept in urban water-infrastructure planning and design. Decentralized systems use locally available water sources at the local level and minimize the need for long-distance water transport. Locally available water resources include captured rainwater (rooftop and other impervious surface runoff), and reclaimed greywater and black water. Advanced small-scale water treatment technologies allow decentralized systems to function as standalone water-infrastructure in appropriate locations such as shopping centers, high-rise buildings, hotels, and college campus buildings. Decentralized systems can also provide water to urban agricultural plots and green roofs for food production in inner cities, college campuses and other appropriate locations. Expected benefits include water and energy conservation, lower food costs and local job creation. Futuristic green water-infrastructure may incorporate decentralized and locally available renewable energy resources, such as solar energy for small-scale water treatment and delivery in urban areas.

Conventional urban water-infrastructure has been practiced for over a hundred years. The shift toward decentralized water-infrastructure can be a significant challenge to water planners, local governments and water utilities. There are technical, policy and perception hurdles to overcome. The first step is to keep an open mind in rethinking water-infrastructure and the needed transition to green and decentralized infrastructure. New urban development and retrofit projects provide excellent opportunities to implement innovative 21st century technologies. Also, a significant need exists to upgrade college level curricula to educate future planners, scientists and engineers not only on mechanical aspects of the water-infrastructure but also understanding the interconnectedness and oneness of natural and engineered water systems. Let's think of current water crises and challenges as opportunities to implement innovative 21st century concepts and solutions.

2.5 Water Management Challenges in Virginia's Karst Terrain

In Virginia, the landscape along the beautiful Valley and Ridge (Physiographic) Region, 28–29 counties in the western part of the state along West Virginia and Tennessee border, is dotted with caves and caverns, springs, sinking streams and sinkholes. Technically this type of terrain is known as karst landscape. Karst is a unique landscape and topography which is formed due to long-term (millions of years) water percolation and action causing dissolution and erosion of carbonate rocks such as limestone and dolomite. According to Wikipedia, the word *karst* is of Mediterranean origin, a form of the language first used in Slavic languages (1689), entered German scientific literature in late 19th century, and it was then adapted to the English language.

Karst landscape is worldwide phenomenon existing on all continents. Depending on the specific type of karst landscape, it can either be easily observed or hidden from the ground surface. Extensive caves and caverns and underground rivers and groundwater aquifers may exist beneath the surface. According to the U.S. Geological Survey (USGS), in the U.S. major karst landscapes exist in more than 20 states and makes up over 20 percent of the total U.S. landscape. According to the Virginia Department of Conservation and Recreation (DCR), it's estimated that 18 percent of the land area in Virginia is karst landscape and the dominant karst region in Virginia is the Valley and Ridge Region.

Caves are the most prominent and permanent feature of karst areas and are extensively documented in ancient writings. For example, the first mention of a cave in Scriptures occurs in Genesis 19:30 (Lot and daughters lived in a cave). Caves can be very small (a person can hardly get through) or very large. In 2001, the Tora Bora cave complex in eastern Afghanistan was a major terrorist headquarters. On a positive note, extended or large caves known as caverns are beautiful tourist attraction sites. In Virginia, there are Luray Caverns, Shenandoah Caverns, Dixie Caverns and a few others.

Sinkholes are the most complex feature of karst landscapes and indicate bedrock instability. Sinkholes can occur through long-term natural formation or human activity. Sinkholes that occur naturally usually form by the slow downward and long-term dissolution of carbonate rock or through bedrock collapse. Human activity related sinkholes can be triggered by simple alterations in the local hydrology (water movement), e.g., from landscaping a yard. As a matter of fact, the writing of this column was inspired by a friend's observation of the occurrence of a sinkhole in a Blacksburg neighborhood. Inadequate drainage along highways and increased runoff from pavements can also be sources of sinkhole development. Most human induced sinkholes are caused by the lowering of the groundwater table below the soil interface. For example, dewatering an underground limestone quarry for the purposes of water supply can lower groundwater table. In Virginia, there have been several instances of sudden sinkhole development and road collapse along I-81. According to the USGS, in the U.S., the most damage from sinkholes typically occurs in Florida, Texas, Alabama, Missouri, Kentucky, Tennessee and Pennsylvania.

A sinking stream is a flowing surface stream that disappears by flowing underground usually into a cave or sinkhole. Sometimes, the stream reappears on the ground surface as a spring. The general perception that spring water is safe is not true in karst landscapes. A sinking stream which reappears as a spring may have already be exposed to above the ground pollution source. Studies in Scott County, southwest Virginia, proved this possibility. Using dye tracing techniques, we found out that the source of biologically dead (no fish and other living organisms) spring water which flowed as a stream was a sinking stream that was affected by sawdust pollution.

From the water supply perspective, karst landscapes pose significant challenges since karst terrane provides a direct and fast linkage between ground surface and groundwater systems. Unlike normal groundwater aquifers where rainwater from the ground surface infiltrates through a natural soil layer and recharges the groundwater – the soil layer serves as a filtration mechanism that removes many contaminants in water – in karst regions, surface water directly enters underground aquifers via fractures and channels and thus groundwater is highly susceptible to pollution from above the ground activities, such as animal farms and contaminant spills. It is noted that in the U.S., karst aquifers provide 40 percent of the nation's water supply. Despite a recent increase in knowledge about the complex nature of karst landscapes, in Virginia and elsewhere, karst landscape protection and water management remain a major challenge to scientists, engineers, regulatory agencies, and other stakeholders.

2.6 The Human Dimension of Water Management: Learning from Blacksburg History

Land development, population growth and water use are intimately interrelated phenomena. Blacksburg, Virginia is a typical small-town America with the exception that it's the home of Virginia Tech where historical data are preserved extensively. The human dimension refers to an understanding of the societal benefits of land development and water use, but also reflects the negative impacts of land development and population growth on water resources. This understanding can lead us to innovative and futuristic management of land and water resources.

The focus of this article is on Stroubles Creek which originates from three springs in the Town of Blacksburg. Three streams, mostly underground, flows through downtown Blacksburg and merge at the Duck Pond on Virginia Tech campus. Then it flows 9.2 miles and eventually drains into the New River. The New River flows north to the Kanawha River, which in turn flows to the Ohio River. The Ohio River merges into Mississippi River. The Stroubles Creek flow demonstrates that our local activities can impact larger rivers.

The Stroubles Creek area was first settled in 1740 by the Draper's Meadow Community, and of course we must realize that the area was already inhabited by a native population. The three springs provided a vital water source for the community. By 1798, the settled land became the Town of Blacksburg – a sixteen block 38-acre square grid. In 1851, the Town's Methodist community opened the Preston and Olin Institute, a seminary for boys. By 1860, the population of Blacksburg had grown to 460 people. In 1872, the Virginia Agricultural and Mechanical College was established as one of Virginia's Land Grant Universities (the other is Virginia State University). In 1896, the college's name was changed to the Virginia Agricultural and Mechanical College and Polytechnic Institute and later to Virginia Polytechnic Institute and State University or Virginia Tech.

The University became the driving force behind population growth and land development, i.e., conversion of agricultural fields and forest lands to buildings and roads. Blacksburg downtown and parts of the University were built over free flowing ancient streams. Until the 1940s, land development was rather slow. However, after World War II there was a significant increase in university enrolment and town population. From 1950 to 1960, the population increased from 3,358 to 7,070; and from 1960 to 1980, a second population boom, the population increased from 7,070 to 30,638. Before World War II, the area depended on natural springs for its water supplies. After the population boom, major changes in terms of water supply became necessary. It was determined that the New River was the practical water supply source for the town and University. In 1950, a special Act of the legislature created the Blacksburg-Christiansburg-VPI Water Authority which is responsible for drinking water supply in the area. The increased population also resulted in high volumes of sewage from the town and the University. In 1948, a sewage treatment plant was constructed about 4.5 miles downstream from Virginia Tech. The second and the existing wastewater treatment facility went into operation in 1978. Treated wastewater is discharged back into the New River, thus closing the loop.

1937 Aerial Photo of Drillfield and Duck
Pond (Source: Jim Campbell 2009)

Dam below the Duck Pond
(Source: Parece 2010)

University researchers began studying Stroubles Creek water quality as early as 1914. These studies covered different parameters and varying locations. Over time, several activities and incidents have contributed to water pollution in Stroubles Creek. From the 1800s to the 1930s, coal mining wastewater contaminated the lower part of the Stroubles Creek. From 1970 to 1978, chemical waste generated in Virginia Tech's Chemistry Department laboratories was directly discharged into the nearby Duck Pond. Today, the Town of Blacksburg and Virginia Tech are still the major contributors to the creek water quality degradation because of urban stormwater runoff and animal farms adjacent to the Stroubles Creek. In the third decade of 21st century, the town and Virginia Tech depend on energy intensive water supplies pumped from the New River. There is significant need for water and energy conservation measures, and implementing innovative green water-management technologies such as rainwater use, and wastewater reclamation and reuse to mitigate climate change impacts. Acknowledgement: Tammy Parece, Stephanie DiBetitto and Tiffany Sprague contributed to original report titled: The Stroubles Creek Watershed: History of Development and Chronicles of Research.

2.7 Water Management in the Era of Information Technology

Water availability and quality for human uses and ecosystem preservation are significantly affected by rainfall's interaction with land, i.e., forests, agricultural and urban. Watershed protection and water monitoring programs are critical tools for management and protection of water resources. A watershed depicts a geographic boundary, where rainfall collected within the bounded area drains to streams, rivers, and lakes. A large watershed can contain several smaller watersheds and can cross political boundaries, such as towns, counties, cities, states and even countries. For example, the New River watershed, one of the 14 major watersheds in Virginia, is divided between North Carolina (headwaters), Virginia and West Virginia. In Virginia, 32 tributaries flow into the New River (i.e., sub-watersheds of the river) as it flows through the counties of Grayson, Carroll, Wythe, Pulaski, Montgomery and Giles.

Manual water monitoring techniques for water quantity and quality assessment were developed during the late 19th and early 20th centuries. In general, manual or discrete water monitoring is conducted at regular time intervals (biweekly or monthly) and provides a broad view of seasonal changes of water quantity and quality. However, these do not produce sufficient data to capture temporal (relating to time) changes in water flow and quality that occur during episodic events such as major storm events, pollutant spills, and harmful algal blooms. In recent decades, significant advances in water monitoring technologies allow real-time (the actual time during which a rainfall event occurs) and continuous water monitoring during rainfall events, and also allow data collection during times when it is impractical (e.g., night) or dangerous to send sampling crews to the field. While discrete water monitoring only provides a snap shot in time, much like a photograph, continuous water monitoring provides a continuous data on conditions, similar to a movie, which results in a better understanding of environmental processes and the ability to capture data during episodic events. Noted benefits of real-time water monitoring include flood forecasting, drought management, drinking water source protection, ecosystem management, and real-time response to potentially catastrophic events.

Advanced monitoring technologies deploy water quantity/quality sensors, data transfer platforms and remote sensing technologies. Using water level sensors to determine real-time stream/river flow has become a common practice. The U.S. Geological Survey (USGS) has upgraded its network of water level monitoring stations on rivers across the U.S., where at each station real-time water level data is typically recorded at 15–60-minute intervals, stored on data loggers (electronic device that records data over time), and then transmitted to the USGS offices every 1 to 4 hours via satellite, telephone, and/or radio telemetry.

The USGS National Water Information Web Interface is in public domain. For example, a citizen can observe daily recorded water level data for the New River at the Radford, Virginia station (USGS Site: 03171000) https://waterdata.usgs.gov/nwis/uv?03171000. Water quality sensors are an evolving technology that gradually enables monitoring of a broad spectrum of contaminants in water, and presently are used for diagnostic purposes to detect symptoms of possible water quality degradation. Currently, commonly monitored parameters include water temperature, pH and conductivity, dissolved oxygen, and turbidity.

A revolution is underway in remote land and water monitoring technologies from using drones to satellites and airborne remote sensors that can monitor land use and water from above the Earth's surface. Satellite imageries are used to observe land use changes such as urbanization, deforestation and desertification; detect water in semi-arid and arid zones; groundwater movement and storage; assessing water quality parameters in inland waters such as lakes, streams, rivers, reservoirs and ponds. For example, in the Midwest U.S., historic and recent Landsat (a series of U.S. land-observation satellites) water clarity assessments have been conducted on more than 20,000 lakes to investigate spatial and temporal patterns and explore factors that affect lake water quality.

It's important to note that small tributaries are major sources of water pollution in rivers and lakes, and protection of small watersheds should be a high priority. In 2010, our research team at Virginia Tech established the Learning Enhanced Watershed Assessment System (LEWAS) real-time water monitoring laboratory on Virginia Tech campus, upstream of the Duck Pond, for the Webb Branch of the Stroubles Creek – a tributary of the New River. LEWAS has deployed sensors to measure water quality and quantity data including flow rate, depth, pH, dissolved oxygen, conductivity, and temperature. This data can be used as an indicator of stream health for an on-campus impaired stream in real-time, and provide an incentive to clean up and restore the water quality of the Duck Pond – a major recreational spot at Virginia Tech and Town of Blacksburg.

2.8 Water Management in the Era of Cybersecurity and Artificial Intelligence

Late 20th and early 21st century revolutions in computerization and internet technologies affect all aspects of our lives. The impact is very dramatic on water security and water management, which significantly affects human life and ecosystems. A precious piece discussed smart technologies such as using sensors and drones combined with wireless internet technologies that can ensure the security of the nation's water resources; transmitting fast and immediate distribution of real-time water quantity/quality information to governmental agencies and other stakeholders for speedy responses that could alleviate/mitigate natural disasters, such as floods and droughts, and artificial disasters such as chemical spills.

Other examples of smart technologies developed in the 20th century include "smart water meter" and pipeline leak detection technologies. A "smart water meter" automatically measures and transmits potable water consumption for billing purposes. Pipeline water leak is a critical issue. The American Society of Civil Engineers (ASCE 2017) has estimated that 240,000 water main breaks occur yearly in the U.S., with 2 trillion gallons of treated and energy intensive drinking water wasted. Conventional leak detection method was to observe visible evidence of leaking, opening hydrants, and digging up pipes. An example of smart technology for leak detection uses acoustic signals (detects leak noise) which determines the condition of underground water pipes for quick leak detection and repair.

Beyond smart technology, research and technology advances in the 21st century have led us to the era of cybersecurity and artificial intelligence with important applications in water security and management. Cybersecurity refers to the preventative techniques used to protect the integrity of networks, programs and data from attack, damage, or unauthorized access. Artificial intelligence (AI) refers to creation and use of digital data, and using intelligent machines and robots.

The 9/11 terrorism act of 2001, triggered the evaluation of water infrastructure safety in the United States, among other concerns. While advances in water sector computerization provide significant advantages, the process has also exposed the water system's vulnerabilities to hacking and terrorism, therefore, making cybersecurity a top priority for the water sector. A recent American Water Works Association (AWWA) Report 'Cybersecurity risk and responsibility in the water sector (AWWA 2018)' states that *'Government intelligence confirms the water and wastewater sector is under a direct threat as part of a foreign government's multi-stage intrusion campaign, and individual criminal actors and groups threaten the security of our nation's water and wastewater systems' operations and data.'* The report states that *'Attacks causing contamination, operational malfunction, and service outages could result in illness and casualties, compromise emergency response by firefighters and healthcare workers, and negatively impact transportation systems and food supply.'* The AWWA report presents several examples of confirmed water sector attacks in the United States. According to the report, in one case, cybercriminals exploited antiquated computer systems to gain access to valve and flow operations, and were able to manipulate the water flow and amount of chemicals used to treat the water. In another case, attackers exploited a vulnerability to identify an unprotected computer that controlled sluice gates and other functions of a dam. To meet the cybersecurity challenge, America's Water Infrastructure Act (AWIA) was signed into law on October 23, 2018. Under the AWIA, water utilities have to conduct risk and resilience assessments and revise their emergency response plans to address cybersecurity.

Application of artificial intelligence (AI) in water security and management is very promising and exciting. Robots have already found important applications, such as in water sampling and analysis, and leak detection in water networks. An important aspect of AI is the capability to create and share "digital data" using existing databases and continuously adapt to changes when new data becomes available. This AI function, digital solution, has significant application in water resource management. Using digital water data enable water managers and governmental agencies to plan efficient and sustainable water resource management systems and to build efficient and sustainable water infrastructure. Sustainable water systems will result in water and energy use efficiency, mitigate climate change impact on water and energy resources, and, in some ways, complement cybersecurity in the water sector.

2.9 Virtual Water Consumption: The Invisible Wasted Waters

In the United States, June–August is the moving time as many people are transferring to new jobs and new locations. The moving phenomena, and its aftermath, is particularly visible in university towns such as Blacksburg when students are graduating or returning home for summer, and as a friend recently observed, you can see many commodities such as furniture at waste dumping sites around town. What is the implication of this situation in terms of water consumption?

In general, two categories of water use include direct and indirect uses. Direct uses of water are rather transparent and include daily domestic uses of water such as water used for drinking, bathing, cooking, laundry, gardening and lawn watering. In a larger scale, direct uses can include activities such as farm irrigation, animal production, energy extraction and electricity generation. Direct use of water is metered. Customers pay for direct water use and often think of water conservation measures. Less known to most consumers is the indirect uses of water and the fact that water is a necessary resource for producing all types of commodities – food, alcoholic and non-alcoholic beverages, clothing, furniture, cars, computers, cell phones – just a small example of many other products. The term "virtual water" expresses the amount of hidden water used to produce commodities and is also called a water footprint. Less is known about the water footprint of all manufactured products such as furniture. A few examples of known virtual water use include producing and processing enough coffee beans for 1 cup of coffee (37 gallons); raising and processing enough beef for a ½ pound hamburger (630 gallons); producing one pair of cotton jeans (2,100 gallons); making a smart phone (3,200 gallons); manufacturing a car (13,700–21,900 gallons).

A significant feature of the manufacturing process and virtual water use is that large volume of water move with commodities away from the production location to consumer location; across regions, states, and/or abroad. For example, U.S. exports virtual water abroad through trade of agricultural products and imports virtual water via commodities such as cloths, cell phones and other products. As noted in a previous column, fair trade from a virtual water perspective is a complicated international issue. For example, Grace Communications Foundation describes the process to produce a smart phone as follows: '*Phones are composed of many pieces created in multiple steps, and each step consumes water. Numerous resources, materials and parts go into smartphone manufacturing, including rare earth metals (e.g., lithium), tin, glass and plastics. The supply chains for these materials stretch around the world to places like Indonesia, the Philippines and China. Production might include steps like mining for precious metals, creating synthetic chemicals for glue and plastic and assembling and packaging. Collectively, the water associated with each step adds up to the water footprint.*' It should be noted that the concept of virtual water use similarly applies to virtual energy use as large volumes of water is used to produce various forms of energy used in manufactured commodities.

Water and energy use efficiency are primary factors in economic development and sustaining a healthy standard of living. Thus, recycling of commodities, where possible, becomes critical from water and energy management perspective. There is a significant need to increase citizen awareness regarding recycling and its connection to virtual water and energy use and wastage. In a futuristic society, there should be greater focus on policies, technologies and citizen behavior that reduces wastage of virtual water and energy. For example, significant reduction in virtual water and energy wastage can be achieved not only by recycling but also direct uses of locally available alternative water resources (such as rainwater) and renewable energy resources (solar, wind, other) for commodity manufacturing (and food production) at the local level.

2.10 The Promise of Rainwater: From Mountaintops to the Shining Sea

In late 1990s, while studying drinking water problems in the coalfield counties of southwest Virginia, findings were alarming. Generally, Virginia coalfield geology does not support plentiful groundwater supplies. The area's primary subsurface aquifers are coal seams that are limited in their ability to supply groundwater. In areas of extensive underground coal-mining, these aquifers have been disrupted, and conventional wells are not possible. Providing water supplies to these communities through a public water distribution system is generally cost prohibitive because of the rough and elevated terrain. To meet potable water needs in coalfield counties, rainwater capture can be considered a valuable alternate water source for domestic and other consumption. As a matter of fact, in coalfield communities of southwest Virginia, rooftop rainwater collection in conjunction with cistern storage has been practiced for many years, particularly in small mountaintop communities where lifting potable water is very costly. Recent advances in rainwater harvesting technology – defined as rainwater capture and use – provide a promising approach for securing safe and adequate water in coalfield communities and should be considered a significant component of economic development in pursuit of implementing green technologies in the coalfield counties.

Lack of public water supplies in rural and isolated communities is not unique to coalfields of southwest Virginia. Nearly, 50 million people in the U.S., including rural communities in Virginia, depend on private water wells. Extending public water to rural and small communities is costly – it's estimated that extending public water pipes to an isolated small community cost $300 per foot of pipeline. Groundwater resources in rural communities are highly susceptible to pollution from surface activities such as farming and animal production, and natural geologic conditions such as karst formations – known karst characteristics are sinkholes, caves and sinking streams – allow direct water and contaminant percolation from ground surface to groundwater aquifer without the advantage of natural soil filtration. There are 28 counties in Virginia, mostly rural, which depend on karst aquifers for their water supply. Well water testing across the U.S. and Virginia has provided sufficient evidence of groundwater pollution in rural communities. Emerging contaminants such as hormones and pharmaceuticals in source water pose a significant threat to public health in rural communities. Fracking to produce natural gas, and its uncertain impact on groundwater quality, is also becoming a national concern. Rural communities can consider rainwater harvesting as an attractive alternate water source because of its better quality as compared to groundwater.

At present, over 80 percent of the U.S. population live in metropolitan and non-metropolitan urban areas and are served by public water utilities. In general, about two-third of the energy intensive potable water from utilities is consumed for non-potable uses such as flushing toilets, washing cars and landscape irrigation. Rooftop rainwater harvesting provides a significant opportunity for potable water saving. Potential outdoor uses of rainwater include lawn irrigation, fountains, gardening, and car washing. Indoor uses of rainwater include flushing toilets, laundry and cooling. Advances in small-scale water treatment technologies have enhanced the potential for using rooftop rainwater for potable purposes in both residential and commercial buildings.

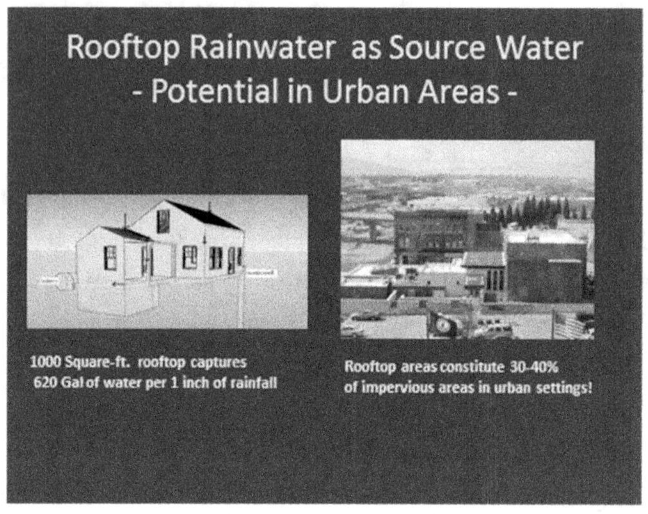

Source: Author

In an urban scale, potential benefits of rooftop rainwater harvesting include: reduced demand for utility water, reduced stormwater runoff volume to drainage network, flood control, drought management, and reduced energy use attributed to lower utility water use. Rooftop rainwater sharing from neighboring buildings in an urban area can supply adequate water to community gardens and urban farms enhancing food security and possibility for job creation, and enhance urban aesthetics – supply water to protect instream flow and ecosystem, and create community recreational lakes. Most importantly, direct well injection of rooftop rainwater can be considered a significant tool for drought management where half-empty aquifers are prevalent. Groundwater preservation is most critical in coastal urban areas. At present, about 50 percent of the U.S. population lives in coastal cities where high water demand has caused significant drop in groundwater table resulting in saltwater intrusion in coastal aquifers including Eastern Virginia. Direct well injection of captured rainwater in coastal aquifers of Eastern Virginia can create a "water curtain" between freshwater and saltwater interface and prevent saltwater intrusion in these aquifers.

In conclusion, there are significant opportunities and benefits for municipalities and other stakeholders to incorporate rainwater harvesting in developing sustainable water management strategies for localities across Virginia – from mountaintops to the ocean.

2.11 The Vision for a Pipe-Less Society

Urban sprawl, the uncontrolled expansion of urban areas, has significantly affected the characteristics of our rural environments due to substantial losses of agricultural and forest lands to urban development. According to the USDA Forest Service, *'every day, America loses more than 4,000 acres of open space to development; that's more than 3 acres per minute.'* The process of land development typically involves alteration of natural land and water systems – land grading, removing vegetation, soil compaction and creating paved and impervious surfaces, i.e., parking lots, roads, rooftops. As a consequence, conventional land development produces high volume of stormwater runoff and increased flooding potential, less infiltration and natural groundwater recharge. Other consequences of urban sprawl include the need for the expansion of water/wastewater/stormwater infrastructure which mainly depend on expanding pipelines; and increased energy demand for operating energy intensive centralized water infrastructure.

According to Natural Resources Defense Council (NRDC), each year an estimated 10 trillion gallons of urban stormwater runoff enters into United States waterways. And according to our estimate, approximately 8650 million gallons of stormwater runoff enters waterways from paved areas within the City of Roanoke. Urban stormwater runoff is a valuable wasted resource and contaminants associated with urban stormwater runoff, i.e., metals – zinc, copper and lead; pesticides and fungicides; oil and grease; sediment and nutrients (N, P); and pathogens are the primary cause of water quality and ecosystem degradation in our streams/rivers and lakes.

Major challenges we face in the 21st century include emerging contaminants (for example, pharmaceutical and hormones) in surface and groundwater; half-empty groundwater aquifers across the U.S., including Virginia; potable water leakage and contamination via pipelines; chemicals and emerging contaminants in our drinking water; man-made flood and drought in urban areas; high energy demand in water infrastructure; consequences of climate change such as sea level rise; and cyber security of centralized water and power infrastructure. This year's theme for World Water Day celebrated on 22 March is "Nature for Water" and explores nature-based solutions to the water challenges faced by communities throughout the world. The vision for a pipe-less society, briefly described below, provide nature-based solutions which can be attained at the local level.

The vision for a pipe-less society promotes the integration of natural and engineered systems to meet water and energy demand and protect ecosystem in our urban and rural environments. The concept of a pipe-less society is based on minimizing dependency and transport of water/wastewater/stormwater and energy via pipelines and maximizing the development and use of locally available water and renewable energy resources with no or fewer pipelines. A pipe-less society can be achieved by implementation of small-scale decentralized water and renewable energy infrastructure. Locally available water sources may include rainwater, stormwater runoff, groundwater, wastewater, and brackish/saltwater. Locally available energy resources may include solar, wind, micro-hydro, geo-thermal, biomass and other. Decentralized water and renewable energy infrastructure will also assure a higher level of cyber security compared to huge centralized water and energy infrastructure.

Some examples of small-scale and decentralized water infrastructure are green technologies that include rainwater harvesting – a technology that collects and stores rainwater from the rooftops and impervious land surfaces for various indoor and outdoor uses, such as potable water, flushing toilets, fountains, landscape irrigation, urban agriculture and community gardens, swimming pools and recreation ponds; low impact development such as rain gardens which are typically composed of a mix of components such as a buffer strip, vegetation and pond areas; daylighted urban streams, i.e., removing streams from underground pipes; small-scale onsite wastewater treatment which relies on natural process and/or mechanical components to collect, treat, and dispose or reclaim wastewater; and underground rainwater/stormwater storage and recovery systems for local use and groundwater preservation. Small-scale and decentralized renewable energy infrastructures are evolving technologies that are rapidly developing for various applications and present great promise and potential for integration with decentralized water infrastructure and other uses at the local level.

As we proceed to plan and design self-sufficient urban/sub-urban and rural water and energy infrastructure, it is essential to consider the vision for a pipe-less society into retrofits and new developments, and encourage local governments to incorporate green technologies in zoning ordinances and building codes. Local water and energy problems can be best addressed at the local level.

2.12 Stream daylighting: The 21st Century Green Water-Infrastructure

Twenty-first century America is approaching a turning point in its approach to urban stormwater management, i.e., stormwater runoff collection and drainage from imperious land surfaces. The 20th century engineering that made rapid land development possible is often failing and creating a host of problems. Cracked and collapsing pipes causes major urban floods as undersized culverts fail to handle the stormwater runoff generated during heavy rainfall events on impervious surface areas such as roads and parking lots. Today, keeping natural streams underground to facilitate land development remains a common practice, and many municipalities continue to replace and/or expand the underground piped stormwater drainage system. Fortunately, current trends in environmental awareness and stewardship are making it possible and imperative to imagine and build sustainable water management systems for our cities and to preserve rivers, streams, creeks and natural ecosystem. Movements toward implementing "green water-infrastructure" such as low impact development (LID) and environmental best management practices (BMPs) are gaining ground in public debate, policy making, and land use management and planning. Stream daylighting is a LID approach which is trending toward increased implementation.

The word "daylighting" is often unfamiliar to most people, who confuse it with bringing daylight into the interior of a room or building. The term daylighting describes stream restoration projects that deliberately expose part or all of a previously covered stream, river or stormwater drainage pipes and restores the underground waterway to open air. Measurable benefits of stream daylighting include improved riparian habitat and water quality along newly created stream banks; reduced flood impacts by increasing storage capacity in comparison with underground culverts; reduced urban "heat island" effect – built up areas that are hotter than nearby rural areas resulting in increased summertime peak energy demand, air conditioning costs, air pollution and greenhouse gas emissions, and heat-related illness and mortality; and adding valuable public open space to dense urban communities which consequently enhance aesthetic environment, increased property values and business investment opportunities.

Our research of several case studies shows that stream daylighting projects have successfully reduced urban flooding and enhanced aesthetic value of built environments. The physical constraints of a given location yield three common stream design options and outcomes – artificial streams, channelized streams and naturalized streams. Artificial streams are usually established in highly built urban environments where little space is available for a meandering and shaded stream bed. Its function is limited to controlling water's flow path without restoring the stream's basic ecological function. Channelized streams are typically found in urban or suburban areas where large vacant parcels facilitate a higher degree of ecological function and aquatic habitat. Naturalized streams are typically found on large pieces of property such as school fields and campuses where greater land area exists to re-establish floodplains, wetlands, ponds, and wider stretches of riparian plantings and forest buffers. Naturalized streams offer the highest degree of ecological function and typically see the largest number of returning fish and insect species within their channels.

Though the first stream daylighting project occurred in 1984 along a section of Strawberry Creek in a Berkeley (California) park, today there is an increasing trend toward implementing stream daylighting projects across the country. Stream daylighting has proved to be a viable green water-infrastructure alternative to traditional stormwater management approach. In appropriate situations, daylighting and open streams are becoming a successful retrofit method for controlling specific types of urban floods and consequent damage that we often observe and experience. Although stream daylighting initial cost is perceived to be high, in many cases long-term daylighting cost can be less than expanding and installing new underground pipes. Stream daylighting and environmental preservation can also result in community warm-glow feeling and emotional reward. Our planners, engineers, and municipalities need to reevaluate high cost and less effective stormwater management projects, and where feasible, consider a stream daylighting option as an alternative approach for stormwater management and flood control. Acknowledgment: Tracy Buchholz Strickland, a landscape architect and former graduate research assistant at Virginia Tech, conducted research on stream daylighting and co-authored a book chapter on this topic.

3. Water and Energy Connections and Link to Climate Change

3.1 Energy Use and Water Resources Impacts

Fossil-fuel energy consumption and carbon dioxide and methane emissions attributed to climate change are widely debated. However, there is less attention to water dependency of energy resources on water resources. According to the USGS 2009 report, on a daily basis, 410 billion gallons of water are withdrawn from the U.S. rivers, lakes and groundwater. Thermoelectric power generation is the largest water user (201 billion gallons/day or 49% of total withdrawal), followed by crop irrigation (128 billion gallons/day or 31% of total withdrawal). Public domestic water consumption was in third place (11 billion gallons/day or 11% of total freshwater withdrawal). Large volume of water is used to extract fossil-fuel energy and to generate thermoelectricity.

At present, fossil fuels contribute more than 80% (petroleum 36.2%, natural gas 29%, coal 16.1%) of energy use in the U.S., while 20% comes from renewable energy sources (nuclear 8.5%, biomass (plant or animal) 4.8%, hydroelectric 2.4%, wind 1.9%, solar 0.5%, geothermal 0.2%).

In coal production, water is used during mining operations, coal transport and storage, coal refining process, dust suppression, and during post-mining activities such as land reclamation and revegetation. Water is a major component of natural gas and oil extraction – large volume of water is withdrawn when gas and oil are extracted but this water is mostly re-injected into the aquifer. Water is also used in oil refinery process.

The greatest threat to water resources is posed by mountaintop surface and strip-mining operations. These operations expose coal seams to be extracted and often dispose the removed overburden (called spoil) in adjacent "valley fills". This result in destruction of natural topography, elimination of natural vegetation, punctured groundwater aquifers and degradation of nearby streams and rivers and associated ecosystems. Acid mine drainage, the movement of highly acidic water, from active or abandoned mines, can seep into stream and rivers and groundwater aquifers, with significant impact on human health and ecosystems. Advances in clean coal technologies such as carbon dioxide capture and sequestration from coal-fired power plants may improve air quality by reducing CO_2 emissions but it does not reduce threats to water resources and ecosystem.

At present, natural gas is promoted as a clean energy source and is gradually replacing coal for electricity generation. Levels of CO_2 emission for coal and natural gas are, respectively 2.117 and 1.314 pounds/kwh of power generated. Compared to coal, natural gas extraction and natural gas-fired power plants are more water efficient. However, recent studies show that significant methane leakage from natural gas wells offsets natural gas advantage over coal to reduce climate change impact. Furthermore, using hydraulic fracturing (fracking) technology to extract natural and shale gas (oil as well) diminishes the advantage of developing natural gas. Fracking typically involves injecting water, sand, and (or) chemicals under high pressure into a bedrock formation via a well. There is strong evidence on fracking impact on groundwater quality and availability.

There is a critical need for using renewable energy technologies that are less water intensive. Existing major renewable energy technologies that generate electricity include nuclear energy and hydropower. Despite advances in technology, nuclear electricity generation is still highly water intensive. Developing hydropower which is the least water intensive of renewable technologies is impeded by land availability to impound water and ecological impacts of dams and impoundments. However, environment friendly decentralized micro-hydropower systems are futuristic technologies in the horizon.

Bioenergy or biomass is a promising renewable energy for generating heat, electricity and transportation fuels (biofuels) and there are many successful projects. One exception is the corn and soy-based biofuel and ethanol production technologies that are highly water intensive due to significant water use in irrigated agricultural fields. Geothermal power technology – energy extracted from the natural heat stored beneath the earth surface by harnessing the steam – is water efficient. However, the availability of geothermal energy is localized and can be considered as a valuable decentralized renewable energy source where available. Solar thermoelectric power is the most recent significant renewable energy source. Solar thermoelectric power generation uses water for steam generation, as a coolant and cleaning purposes.

Water demand for energy production and electricity generation is in competition with potable water demand and food production. There is a significant need for water use efficiency of energy resources. This can be accomplished by adopting a multitude of decentralized and water efficient renewable energy technologies (where locally feasible) such as wind, solar photovoltaics, geothermal, bioenergy, micro-hydro, and other upcoming innovative energy technologies that minimize the impact on water resources and reduce climate change threats.

3.2 Water and Energy Consumption in the Era of Globalization

In our modern world, we use water for domestic purposes such as drinking, bathing, cooking, laundry and watering the lawn. We use energy for lighting, household appliances, cooling and operating our computers. These are transparent and direct uses of water and energy which are metered and directly paid for by consumers. In a larger scale, water is used for activities such as farm irrigation, animal production, energy extraction and electricity generation. Less known is the fact that water and energy are necessary resources for producing all types of commodities – food, alcoholic and non-alcoholic beverages, clothing, furniture, cars, computers, cell phones – just a small example of many other products.

The term "virtual water" expresses the amount of hidden water used to produce commodities (see Section 2.9). A few reported examples of approximate water footprint of some popular products include: 37 gallons to produce and process enough coffee beans for 1 cup of coffee; 630 gallons to raise and process enough beef for one ½ pound hamburger; 2,100 gallons to produce one pair of cotton jeans; 3,200 gallons to make a smart phone; and 13,700–21,900 gallons to produce a car.

Similarly, the term "virtual energy" expresses the amount of hidden energy in manufactured products. Energy footprint of a manufactured product is a critical factor because it's linked to the carbon footprint, the amount of carbon dioxide emission attributable to direct and/or indirect consumption of fossil-fuels (coal, petroleum and natural gas). Similar to water footprint, estimating energy footprint is also very complicated. For example, a study at the University of Sheffield (U.K.) collected energy use data for each step in producing a loaf of bread – growing the wheat (machinery use and irrigation), fertilization, harvesting the crop, transporting wheat grains to the mill, grinding the grains into flour, transporting the flour to a bakery, operating the bakery, packaging a loaf of bread, and transporting the loaf of bread to a retail store.

A significant impact of the manufacturing process is that large amounts of virtual water and energy moves with commodities away from production location to consumer location; across the regions and states, and often traded abroad. For example, U.S. exports virtual water and energy abroad through trade of agricultural products and imports virtual water and energy via commodities such as cloths, cell phones and other products. Fair trade from virtual water and energy consumption perspective is a complicated international issue. For example, Grace Communications Foundation describes the process to produce a smart phone as follows: '*Phones are composed of many pieces created in multiple steps, and each step consumes water. Numerous resources, materials and parts go into smartphone manufacturing, including rare earth metals (e.g., lithium), tin, glass and plastics. The supply chains for these materials stretch around the world to places like Indonesia, the Philippines and China. Production might include steps like mining for precious metals, creating synthetic chemicals for glue and plastic and assembling and packaging. Collectively, the water associated with each step adds up to the blue water footprint.*' Blue water footprint is the amount of surface water and groundwater used directly to produce an item. Similarly, it's obvious that certain amount of energy is used to produce a smart phone.

In the era of globalization, water and energy resources are limiting factors in economic growth, food production, and maintaining a healthy standard of living. There is significant hidden cost not only due to the virtual water and energy use in commodities but also direct cost associated with water and energy transport and environmental protection (for example, natural gas pipelines). Water and energy use efficiency, primary factors in economic development and sustaining healthy societies, can be enhanced by developing alternative water and energy resources for domestic and manufacturing uses, and food production at the local level. Alternative water resources include rainwater, stormwater runoff and wastewater which are currently wasted waters. Locally available alternative energy resources include, solar, wind, biomass, geothermal and other. In the long-term, using locally available alternative water and energy resources will reduce overall water and energy costs and sustain a healthy environment. In a futuristic society, there should be greater focus on policies and technologies that support using alternative water and energy resources at the local level for multi-uses.

3.3 Carbon Footprint of Community Water Consumption

A carbon footprint can be defined several ways. Basically, it refers to the amount of carbon dioxide (CO_2) emissions attributable to direct and/or indirect consumption of fossil-fuel (coal, petroleum and natural gas) based energy to sustain modern human activities. Levels of CO_2 emissions for coal, petroleum and natural gas are, respectively, 0.960, 0.868 and 0.596 kg/kWh of power generated. Carbon footprint is a yardstick that determines climate change impact from human activities, and is used to propose mitigation strategies – practices that directly or indirectly reduce fossil-fuel based electricity use and, consequently, reduce CO_2 emissions.

Energy is used in all aspects of modern world – industrial manufacturing, food production, transportation, construction, water consumption, and other activities. For example, significant amounts of fossil-fuel based electricity is used in pressurized sprinkler irrigation systems for crop production. However, activities such as food production are rather complex because of the hidden and indirect uses of energy. A recent study at the University of Sheffield (U.K.) collected energy use data for each step in producing a loaf of bread – growing the wheat (machinery use and irrigation), fertilization, harvesting the crop, transporting wheat grains to the mill, grinding the grains into flour, transporting the flour to a bakery, operating the bakery, and packaging a loaf of bread.

Water services consume a significant portion of energy resources. According to a Congressional Research Service report, in the U.S., about 56 billion kilowatt hours (kWh) or 4% of total national energy consumption is attributed to water and wastewater services. The American Water Works Association Research Foundation reported energy use for potable water treatment and delivery in the U.S. to be in the range of 0.07–0.92 kWh/m^3, with an estimated average of 0.38 kWh/m^3.

Energy consumption in potable water occurs in three stages, i.e., water source development, water treatment and water delivery. For example, in case of Blacksburg, pumps are used to lift source water uphill 107 meters from the New River intake and transport it to a water treatment facility located about 3.7 kilometers from the intake. After treatment, the water is pumped to a high head (pressure) storage tank and then delivered to the Town of Blacksburg, about 13 kilometers from the water treatment facility, by using a booster pump station. In general, energy demand for water treatment is lower than water transportation demand.

In Blacksburg, about two-thirds of the energy use is attributable to pumping and delivery, and one-third attributable to water treatment. In the Town of Blacksburg, we estimated the energy use for water treatment and delivery as 0.44 kWh/m^3. Since approximately 10 percent of electricity use in Blacksburg originates from hydropower (a renewable energy source), estimated total energy-use attributed solely to fossil-fuel based electricity is 0.40 kWh/m^3. For a single academic building on Virginia Tech campus with measured annual water consumption of 5,377 m^3 (1,420,700 gallons), estimated electricity use attributed to water consumption is 2,136 kWh, and the calculated carbon footprint of water consumption for that building is 4,521 kg/year. The Virginia Tech main campus in Blacksburg resembles a microcosm of a high-density urban area.

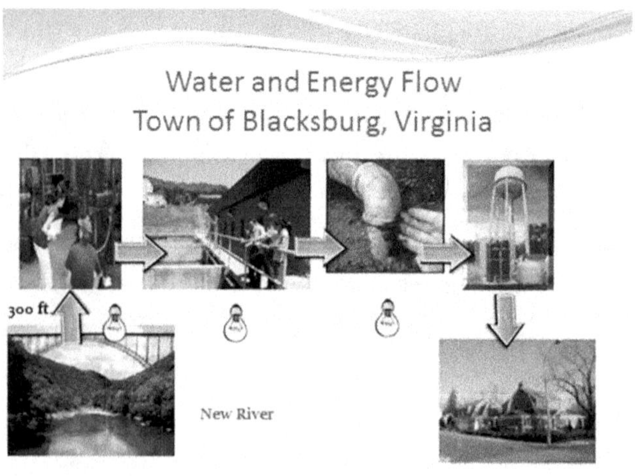

Source: Author

The energy demand for water infrastructure is projected to increase by approximately 30% over the next decades. Mitigation strategies to reduce carbon footprint of water consumption include in-building water conservation such as using less water (behavior change), and adapting water saving fixtures such as low volume showers and toilets, water/energy efficient washing machines, dishwashers and water heating devices, and implementing decentralized water infrastructure such as rainwater capture and use which reduces energy consumption for water delivery via pressurized water pipes. Energy use efficiency in water treatment plants and integration of renewable energy sources in water treatment are evolving technologies. Overall, there is a need for implementing policy options and financial incentives that encourage water and energy conservation, and conducting educational and outreach programs to increase citizen awareness of the connection between water and energy consumption.

3.4 Climate Change: A Water Scientist's Perspective

Climate change refers to changes in the Earth's climatic factors such as average temperature and global precipitation patterns measured over long periods of time, i.e., decades. Climate change science encompasses climate change causes and effects, future projections of climate change and its potential consequences. Climate change investigative tools include atmospheric science, oceanography, ecology, and fundamental sciences of biology, physics, chemistry, and geology.

Climate change science is not new. In 1896, Swedish scientist Svante Arrhenius calculated the warming power of excess carbon dioxide. He predicted that a warming trend would result if human activities increased carbon dioxide in the atmosphere. Averaged over all land and ocean surfaces, temperatures warmed roughly 1.53°F (0.85°C) from 1880 to 2012, according to the Intergovernmental Panel on Climate Change Report (IPCC) (2013). Increase in the Earth's surface temperature is attributed to atmospheric increase in carbon dioxide, methane, nitrous oxide, aerosols and water vapor, i.e., the greenhouse effect.

Possible causes of climate change include both natural and anthropogenic (human) impacts. In terms of human impacts, at present, there is significant focus on energy (fossil fuel) production and consumption as the major cause of climate change and regulatory requirements to control gas emissions. However, deforestation is another significant cause of climate change. Forests store carbon, sequestering it from the atmosphere. Worldwide deforestation, caused by timber industry and accelerated urbanization, contributes to increasing carbon dioxide levels (from loss of sequestration ability and burning of logged trees) and consequently the global warming process.

Measurable impacts of climate change include polar ice melt, change in global ice thickness, sea level rise, and increase in ocean acidity. For example, according to a National Council Report (NCR), the average annual extent of Arctic Sea ice has dropped by roughly 10% per decade since satellite monitoring began in 1978. The NCR report states that since 1870, the global average sea level has risen by 0.66 ft. Two-thirds of sea level rise is attributed to ice melt and one-third attributed to ocean expansion due to warming of oceans. It is estimated that oceans have absorbed between one-quarter and one-third of the excess carbon dioxide caused by human impact including deforestation, but as oceans warm, they lose their capacity for carbon sequestration.

Impact of climate change on water resources and its consequences is the most critical issue of our time. These include changes in weather patterns and consequence depending on geographic location – less snowpack in the mountains and earlier snowmelt, less total annual rainfall, severity and length of droughts, and increases in rain and heavy precipitation events; sea level rise and consequences – flooding of coastal cities and encroachment of saline waters into freshwater aquifers; increased oceans' acidity and consequences – potentially significant impact on marine ecosystems, as well as the health of coral reefs, shellfish and fisheries.

Undated USGS Photo (1885?) 2016

New Point Comfort Lighthouse, Mathews (Chesapeake Bay), Virginia – constructed in 1805 Shoreline has moved 0.5 miles (Source: Wikipedia)

There are two approaches for coping with climate change (IPCC). "Mitigations" are actions taken to limit the degree of future climate change – such as practices that directly or indirectly reduce carbon dioxide emissions to the atmosphere. "Adaptations" are actions taken to cope with current and anticipated problems – actions taken to adjust to effects of climate change as the climate change is already underway – such as sea level rise. Mitigation strategies include energy conservation/energy use efficiency and renewable energy use in utilities, industry, and infrastructure (housing and transportation). Adaptation strategies include prepare to cope with droughts and floods, adapt agricultural production methods to changing climate conditions, develop appropriate policies with a goal to minimize adverse effects of climate change on the economy, on public health and on the quality of the environment.

Unfortunately, with extreme polarized political environment, coping with climate change has become a political movement rather than a scientific and rational movement. While it's important to take gradual steps toward developing mitigation strategies, with the realization that the occurring consequences of climate change are irreversible, it's critical that coping strategies and resources should be more focused on adaptation strategies.

3.5 The Dilemma of Decarbonization

The drive to mitigate climate change has introduced the buzzword decarbonization. Basically, the aim of decarbonization is to reduce operational (direct) and embodied (hidden) global greenhouse gas (GHG) emissions to the atmosphere attributed to human activities. Carbon dioxide (CO_2) is considered the major GHG contributing to global warming. About 65% of atmospheric CO_2 increase is attributed to burning of fossil fuels (coal, natural gas, and gasoline/diesel). In addition, about 11% of the CO_2 increase is attributed to changes in physical and biological characteristics of the land surface such as deforestation, intensive agriculture and urbanization. Other critical GHGs are methane (16%), nitrous oxide (6%) and F-gases – chlorofluorocarbons from refrigerants – (2%). The percentage of CO_2 and methane contribution to atmosphere and their comparative global warming potentials is continuously changing as more scientific data becomes available.

In the US, GHG emissions (operational and embodied) from burning of fossil fuels are distributed across several economic sectors (EPA 2019): electricity generation (28%), agriculture (9%), industry (22%), transportation (29%), residential and commercial (12%). According to the Center for Energy and Climate Solutions, about 65% of electricity generation in the US depends on fossil-fuels (35% natural gas, 30% coal). And according to US Energy Information Administration (EIA) in first 8 months of 2019, renewable energy sources (not including hydro and nuclear power) accounted for 11.4% of US electricity generation: wind (6.94%), solar (2.7%), biomass (1.4%) and geothermal (0.4%). While the gradual decarbonization of power generation plants, i.e., switching to renewable energy resources is an ideal approach, a sudden shift to renewables is not technologically feasible, is considered impractical and cost prohibitive with significant repercussion on other economic sectors because of their dependence on electricity. According to a recent Forbes report, renewable energy sources will require $14 trillion of investment, and could deliver around 80% of global power by 2050.

Effective decarbonization depends on evolving technologies and human behavior. Evolving technologies include but not limited to the design of energy efficient industries, utilities, vehicles, buildings and other infrastructure (embodied carbon footprint), heating/cooling system, and electric appliances and fixtures. During the past few decades, significant progress has been made in the arena of energy use efficiency and is steadily improving.

For example, research shows that using wood instead of steel and concrete to construct high-rise buildings is technically possible and can reduce embodied carbon footprint of built environments. The second evolving technologies relate to design of cost-effective and efficient renewable energy technologies (solar, wind, hydro, tidal, wave, geothermal, biomass, and other). For example, evolving battery technologies for storage of intermittent renewable energy resources, such as solar and wind, are critical for making the shift toward using more renewables. The third component of evolving technologies is the design of smart and distributed (decentralized) energy/electricity grid that automatically integrates various locally available renewable energy resources and also limits the need for long-distance electricity transmission lines.

Human behavior can be characterized as institutional behavior and personal/individual behavior. Institutional behavior is complex. It's influenced by the current state of knowledge, and regulations which are mostly based on the state of knowledge. Advances in technology influences adaptation of new regulation but the process is tedious. Even small-scale changes in institutionalized environmental management, such as building codes and land development to implement energy use efficiency, require significant effort on the part of local and state governments. In our democratic society, policy making is a significant challenge since it's strongly influenced by a market economy and advances in technology. In contrast, personal behavior is an individual responsibility which can significantly impact local and global environment. GHG emissions and climate change are significantly affected by our daily activities. Our existing institutionalized "culture of waste" is the result of collective human behavior which is the sum of individual behaviors. It's said that "small drops make the river", thus to achieve decarbonization goal, we should practice conservation in all aspects of our daily life – energy, water, food, and use of all manufactured products. This is an achievable goal possible with citizen education, and that we should pursue.

3.6 Environmental Values in the Era of Climate Change

Human thinking and values have evolved over time – sometimes within generations. Evolving environmental values are mostly based on advances in science and technology. For example, as noted alter in article 'Coal Production: Panacea or Plague?' large-scale coal mining commenced during the 18th century Industrial Revolution and coal became the primary energy source for major industries such as steel production and electricity generation. The Industrial Revolution resulted in the emergence of high population urban centers and an agricultural sector which demanded more water and energy. As a result, uncontrolled volumes of contaminated domestic, industrial and agricultural wastes were discharged into environment which caused significant water, air and soil pollution. The environmental values of the 18th century were dictated by the 18th century state-of-knowledge and did not foresee unintended consequences on human health and ecosystem degradation which continued to the 20th century.

America was awakened to its environmental problems and its impacts in 1960s. The U.S. Environmental Protection Agency was established in 1970. Major environmental laws such as The Clean Air Act (1970), The Clean Water Act (1972) and The Safe Drinking Water Act (1974) were enacted to protect human health and ecosystems. Environmental regulations are based on scientific state-of-knowledge and are often amended and upgraded (for example, drinking water standards) as new science-based information becomes available. During past 50 years, significant progress has been achieved in protecting the environment. However, in the 21st century, the emergence of climate change poses as a new challenge. Environmental problems noted above are clearly caused by human activities and problems are addressed in accordance to environmental regulations of the affected region and country. Climate change challenge is unique because is triggered by human activities around the world and it affects the atmosphere with global impacts. The controversial nature of climate change noted below presents significant challenge for coping with climate change.

First, most scientists agree that climate change is occurring. However, the cause/or the extent of the cause of climate change that could be natural (solar effect) or man-made is not convincingly settled among climate scientists. Second, climate change studies include: (1) long-term observation, i.e., increasing global temperature and atmospheric carbon dioxide concentration, sea level rise attributed to warming ocean temperatures and ice melting; and (2) computer models which predict future trends on climate change consequences. There is significant scientific confidence in observed data which are based on advanced monitoring and analytical tools. However, there is less confidence in predictive climate models. Complex climate models predicting future consequences cannot be verified and there are contradictory research results regarding the accuracy of climate models.

Regardless of the controversial nature of climate change as noted, there is general agreement that climate change is already occurring. Therefore, there is need to consider ways to cope with climate change. First, it's important to recognize that energy, economy and environment (3EEEs) are strongly intertwined; we cannot delink the 3EEEs, knowing that an effective strategy is only possible under a sound and healthy economic system based on rational thinking and logical solutions. Second, we should recognize that gas emissions are only partially responsible for the man-made component of climate change (for example, think worldwide deforestation). Third, we should recognize that curbing global gas emissions even to zero (which is not possible) may slow down global warming to some extent but it cannot reverse climate change to an ideal condition. Therefore, in the immediate future, adaptation to climate change should be our first line of defense.

Finally, the imminent death of the planet Earth, if we do not take drastic measures, is not a scientifically based statement. That is a wrong message which causes societal hopelessness and panic as evidenced by youth protests around the world. Our environmental values and message related to climate change mitigations should be based on proven science, practicality and American ingenuity. We should focus first at our local level activities and personal behavior. Wasting is the major culprit. Energy use efficiency and conservation practices in all aspects of our lives (energy, water, food and other) should be encouraged and promoted. Also, we should believe in human perseverance and not hopelessness.

3.7 The Promise of Small Hydropower

At present, solar and wind energy are highly promoted as renewable energy technologies – clean technologies in terms of their carbon footprint. However, the most prominent renewable energy source for generating electricity is hydropower. The history of hydropower for generating electricity in the U.S. goes back to late 19th century. In 1893, the first commercial installation of a hydropower plant at the Redlands Power Plant in California allowed electricity to be transmitted long distances for consumer use. At present, hydropower accounts for nearly 9% of the U.S. electric generating capacity.

The traditional hydropower plant depends on a dam and reservoir built on a river. The water released from the reservoir is supplied to a turbine – a rotary engine that uses an electricity generator to convert water energy into electricity. A large hydropower plant can generate electricity greater than 30 megawatts (MW). For example, the power generation capacity of the Hoover Dam built in 1939 and one of the largest in the U.S. is 2,080 MW. In comparison, the power generation capacity of the Claytor Dam (also built in 1939) on the New River in Pulaski County, Virginia is 75 MW. Most reservoirs are designed for multi-purpose uses that include drinking water supply and irrigation water source. In recent decades, two issues have impeded the construction of dams and reservoirs: (1) restrictions attributed to historic value of land and/or high cost to acquire land; and (2) known ecologic impacts of dams such as hindering fish passage. At present, research and technological development is focused on increasing the efficiency of existing hydropower generation plants. However, there is significant hydropower potential in building small and decentralized (distributed) hydropower plants. The electricity generation capacity of small hydropower plant is 100 kW–30 MW. A micro-hydro power plant can generate electricity up to 100 kW. At present, the significant potential of small hydropower generation is untapped.

The most common type of small hydropower plants are built on small streams. As a matter of fact, before rural electrification, micro-hydropower plants were common in small U.S. communities. For example, in the 1940s, Mr. Lucas (a Giles County, Virginia property owner) constructed a small dam and diversion channel on the North Fork Creek that ran through his property at Glen Alton. The channel supplied water to a small turbine that generated electricity. This original power plant was abandoned when the property was connected to the electric grid. The historic site located within the Jefferson National Forest was later donated to the Federal government. In 2013, Virginia Tech research team (Professor Eugene Brown and students) and I in collaboration with the U.S. Forest Service field manager restored the Glen Alton plant for citizen education purposes using a grant from the Appalachian Regional Commission and in-kind support from the USFS.

A small stream hydropower generation plant has fixed and/or limited power generation capacity due to its physical location. However, other emerging technologies can take advantage of the water and energy nexus to generate electricity. For example, electricity can be generated from pressurized pipe flow that can turn a turbine. Portland, Oregon generates electricity from turbines installed in city water pipes – an average of 1,100 megawatt-hours of electricity per year, enough energy to power about 150 homes. High-pressure flows in the pipelines managed by the San Diego Water County Authority allow the generation of electricity through a 4.5 MW turbine generator. These emerging and decentralized hydropower technologies can be expanded to meet electricity demand in both rural and populated urban areas.

To meet electricity demand using renewable energy technologies, there is a need for a holistic approach that integrates several small and decentralized renewable energy resources, for example solar and small hydropower within the same region. Developing small-scale renewable energy resources at the local level will enhance the reliability of energy availability, reduce electricity transmission cost and the dependency on centralized fossil-fuel based energy resources. Regulatory and permitting requirements for renewable energy systems including small hydropower vary in the U.S. according to state and local regulations. In Virginia, the regulatory requirement for small-scale renewable energy development is tied to large electric power (grid) providers. Along with holistic approach, changes in permitting requirements are needed to facilitate efficient development of integrated small-scale and decentralized renewable energy technologies.

3.8 The Shift Towards Using Alternative Water Sources

Water is life! Ancient human communities were established in close proximity of natural water systems, i.e., streams and rivers, lakes, springs and oceans, where water was readily available for human consumption. In modern times, the 18th century Industrial Revolution triggered energy production and use, in the form of fossil fuels (i.e., coal and petroleum) and enabled water transport from source to distant consumers and caused the emergence of high-density population centers. The 20th century witnessed significant modernization in the arena of water resources development and water infrastructure. The 20th century technological accomplishments include building dams and reservoirs, using powered pumps to extract deep groundwater and surface water resources, construction of centralized water treatment plants and water delivery pumps and pipelines to transport clean drinking water to homes and public and commercial buildings, and installing sewer and stormwater pipes to move wastewater and stormwater runoff away from population centers.

Despite advances noted above during the third decade of the 21st century, communities in the U.S. and across the world are being challenged with significant water-related problems such as ecosystem degradation, groundwater depletion, natural and anthropogenic drought (water scarcity), floods, and water borne health issues. These problems are exacerbated by climate change, a phenomenon that is largely accelerated due to human intervention in natural systems since the Industrial Revolution. The impact of climate change on water resources includes, but is not limited to, changes in precipitation patterns and intensity; the severity and length of droughts; sea level rise and associated consequences (e.g., flooding of coastal cities and encroachment of saline waters into freshwater aquifers).

In 2003, the U.S. Department of Interior Bureau of Reclamation, along with the Sandia National Laboratories, created the Desalination and Water Purification Technology Roadmap to meet future water demand in the United States. The 2003 Roadmap correctly anticipated that future water treatment technologies will be expected to treat a variety of impure waters, such as urban stormwater runoff, wastewater treatment plant discharge, and saline water for possible reuse or underground storage. Using alternative water sources – locally available water sources – is now considered a major component of the water management strategies to address complex water management issues in the 21st century. Alternative water sources include rainwater, stormwater runoff, wastewater, salt water and brackish water. Mine cavity water – water stored in old mine cavities and discharged freely through the portals – is a unique alternative water source in the coalfield counties of southwest Virginia.

In recent decades, advances in water treatment technologies have enabled uses of alternative water sources in urban, as well as rural areas, and has been documented through successful case studies across the U.S. and other countries. For example, large scale wastewater reclamation and reuse in Singapore provides significant learning experience. However, at present, alternative water sources are not being sufficiently developed in the U.S. and can still be categorized as "wasted waters" due to several impediments. Unfortunately, from a water management perspective, we still mostly depend on traditional water management technologies (for example, stormwater runoff management) and college curricula which were developed in the 20th century. And policy making, a tedious task, is not keeping up with the advances in science and technology of water management.

There is a significant need for community-based water management strategy and planning which takes into consideration using alternative water sources. Basically, we should consider and plan community development projects with focus on sustainable water management strategies which also integrate using alternative (locally available) renewable energy resources and food production at the local level in the form of various scales of community food production systems. There is a significant need for establishing statewide, regional and local task forces to create guidelines and policy for effective use of alternative water sources. And there is a need for developing water management college level courses that focus on innovations and holistic water management strategies.

3.9 Use of Alternative Water Sources: The Emerging Paradigm

Since the Industrial Revolution, particularly in the 20th century, conventional water sources, i.e., rivers, lakes, and groundwater sources – were developed to meet increasing water demand for public water supplies, and industrial and agricultural use. Dams are built on rivers to create artificial lakes (reservoirs) for various water uses including electricity generation. According to the Virginia Department of Conservation and Recreation in Virginia, there are more than 3,000 reservoirs of various sizes built for many different water uses. It's interesting to note that there are only two natural freshwater lakes in Virginia – Mountain Lake, the site of famed 1987 "Dirty Dancing" movie in Giles County, and Lake Drummond located at the center of the Great Dismal Swamp between Norfolk, Virginia, and Elizabeth City, North Carolina.

Wastewater from houses, public/commercial buildings and industrial activities and stormwater runoff from impervious urban areas are significant byproducts of our modern societies. Generated wastewaters are treated in accordance with regulations and returned to surface waters (rivers and lakes). Urban and agricultural stormwater runoff also finds its way into surface waters. In the 21st century, despite significant progress in wastewater treatment technologies and stormwater runoff control practices, we face major water management problems; anthropogenic drought or water scarcity due to extensive water use in population centers, flooding in urban areas due to increased land development, and surface water quality and ecosystem degradation due to wastewater discharge and stormwater runoff intrusion into surface waters. New approaches are needed to protect river and lake ecosystems. To alleviate existing and emerging problems, innovative approaches include using locally available alternative water sources –

rainwater, stormwater runoff, and wastewater. Rainwater has been used as a drinking water source for generations and still is popular in areas where extending public water lines are cost prohibitive, for example in the mountaintop communities of southwest Virginia. Today advances in rainwater harvesting technology allows commercial and large-scale use of rooftop rainwater for both potable and non-potable purposes.

Uses of stormwater runoff and wastewater as alternative water sources are emerging technologies with significant promise. Advanced water treatment and filtration technologies can produce water acceptable for both potable and non-potable uses. At present, there is a public perception issue particularly related to wastewater reuse and overcoming the "yuk" factor which is considered a major impediment. However, in many localities, we, the people, already unknowingly drink recycled wastewater. Many drinking water plant intakes are located downstream from where treated wastewater is discharged to surface waters. Drinking water plant intakes not impacted by wastewater discharge, such as Carvins Cove – a reservoir and drinking water source for Roanoke, are rare. For example, the City of Richmond and parts of Chesterfield County water supplies use James River water – water flowing downstream that contains treated wastewater from Charlottesville, Lynchburg, Covington discharged into the river. Another notable area in Virginia where recycled wastewater is used for potable purposes is Occoquan Reservoir in northern Virginia, where significant volumes of water flowing into the Occoquan Reservoir originates from an upstream wastewater treatment plant and the drinking water treatment plant intake is located downstream of wastewater treatment plant discharge. Highly advanced water treatment technologies are used to produce drinking water. The Occoquan system is a significant and notable national example of wastewater reuse.

Today, we see national headlines such as 'Toilet to Tap: Wastewater Purification Plant Now Supplying Drinking Water to Oceanside.' Recently, San Diego County in California celebrated the opening of the Oceanside water purification facility which turns wastewater into drinking water using the state-of-the-art technology. According to San Diego officials, the new water treatment plant will eventually create 3 to 5 million gallons per day of local drinking water and could reduce Oceanside's reliance on imported water by an estimated 20 percent.

In terms of advanced and sustainable water management, adapting to climate change is an emerging critical issue. The impact of climate change on water resources includes both drought and flooding problems. Use of alternative water sources including saline waters (seawater and brackish water), not discussed in this column, could alleviate expected problems. Compared to western states, Virginia is considered water rich. However, high and increasing population densities and extensive urbanization in Virginia warrant preparedness in terms of sustainable water management, and should consider planning and design of practices that include use of alternative water sources. Citizen and legislative body awareness on the merits of using alternative water sources should be a high priority as well.

4. Inherited Problems and Scattered Thoughts

4.1 The Dilemma of Pesticides: A 20th Century Problem

My first Op-Ed article, published in the Roanoke Times and World News, titled "Future Farming in America: As alternatives arise, pesticides will be phased out" appeared on December 9, 1990, almost 30 years ago. The time has arrived to look back and see if the optimistic view of a then young scientist has prevailed.

To provide a historic perspective, though chemicals such as arsenic and sulfur were used to kill pests since ancient times, the era of modern synthetic chemicals known as pesticides began in 1930s and greatly advanced in the 1940s. It was symbolized by development and use of the insecticide DDT (*dichloro-diphenyl-trichloroethane*) which was very successful in exterminating mosquitos, carriers of malaria disease. The DDT success prompted the formulation and widespread use of other insecticides, and two other categories of pesticides, namely herbicides for weed control and fungicides for plant disease control. Rachel Carson's book "Silent Spring" published in 1962 brought significant public awareness about the impact of DDT and other pesticides on human health and environment. The Clean Water Act (CWA) of 1972 – reorganized and expanded the Federal Water Pollution Control Act of 1948 – established the basic structure for regulating discharges of pollutants into the waters of the United States and regulating quality standards for surface waters. The CWA is the principal federal law governing pollution in the Nation's surface waters.

Two major governmental actions since 1990 relevant to pesticides are (1) in 1991, U.S. Congress established the USGS National Water-Quality Assessment (NAWQA) Program to monitor and study trends of the Nation's water quality in response to human activities and natural factors; and (2) in 1996, the U.S. Congress passed the Federal Insecticide, Fungicide, and Rodenticide Act (FIFRA). The objective of FIFRA is to protect human health and the environment from unreasonable adverse effects of pesticides. The Act governs the labeling, distribution, sale, and use of pesticides and requires the EPA to regulate the sale and use of pesticides through registration and labeling. According to EPA, pesticides contain "active" and "inert" ingredients. Active ingredients are the chemicals in a pesticide product that act to control the pests. All other ingredients are called "inert ingredients" by federal law and are important for pesticide performance and usability. Pesticide manufacturers must show that using the pesticide 'will not generally cause unreasonable adverse effects on the environment." Pesticides (active and inert ingredients) are associated with risks to human health and adverse effects on water, soil, and air environments.'

According to a 2014 Congressional Research Service (CRS) report, depending on the chemical, possible health effects from overexposure to pesticides include cancer, reproductive or nervous-system disorders, and acute toxicity. According to NAWQA report, for individual pesticides in drinking water, monitoring results are generally good news relative to current water-quality standards and guidelines. Average concentrations of individual pesticides in streams and wells rarely exceeded standards and guidelines established to protect human health. However, for aquatic life and wildlife, NAWQA results indicate problems in many streams, particularly in urban areas, where concentrations of more than one pesticide was found in more than one-half of streams sampled, and concentrations had often approached or exceeded established EPA water-quality guidelines for the protection of aquatic life. Urban use purposes of pesticides include protection of buildings and other structures, weed control in lawns and golf courses, and disease control in home grown vegetables and fruit trees.

According to the USGS and CRS reports, 'important questions remain unanswered about potential risks of pesticide contamination to humans and the environment. Available standards and guidelines do not account for mixtures of pesticides or for pesticide breakdown products, and are based on tests that have assessed a limited range of potential health and ecological effects. Long-term exposure to low-level mixtures of pesticide compounds, punctuated with seasonal pulses of higher concentrations, is the most common pattern of exposure, but the effects of this pattern are not yet well understood.' The CRS report also highlights the conflict between CWA and FIFRA – as a long-standing principle, CWA permits are not required for using FIFRA-approved products.

In conclusion, thirty years later, despite marketing of less (environmentally) persistent pesticides and emerging technologies such as biotechnology applications, and the push for green and organic alternatives, pesticide use and its associated problems will continue for years to come. While we should give credit to scientists, policy makers and concerned citizens for their efforts, still many scientific, managerial, and legislative challenges remain related to continued dilemma of pesticides in our living environment.

4.2 World Citizenry Misrepresented in the Natural Gas Pipeline Argument

Energy, economy and environment are three vital pillars of our modern world. Energy (all forms of it) sustains the economy and our standard of living. These pillars are needs not wants in our daily life. In the 19th century and first part of the 20th century, energy extraction and delivery were the driving force for industrial revolution which laid the foundation for present modern world. In that era, the value of good environment (clean water, clean air, ecosystem, forests and landscape among others) was mostly ignored. The outcome was dirty rivers which caught fire, stripped mountains, destruction of surface and groundwater resources, dirty air and significant impact on citizen health. Also, in that era, private property rights and civil rights were not respected. America was an experiment in democracy with high ideals embedded in the Constitution though there was a huge gap between idealism and reality.

With realization of the negative environmental and health impacts of unrestrained economic development, environmental laws were enacted in the U.S. in the early but mostly later parts of the 20th century. It was recognized that a sick society is not a happy society. It was also recognized that health, happiness and rights of a single individual is as important as the general welfare of the entire society, and for that matter the entire nation. A society and a nation are made of its individual members; all of them are created equal.

Unfortunately, many experts support status quo and justify our past mistakes (unintentional, based on the status of knowledge), and does not provide a vision for moving us forward to a new era of prosperity. The pipeline debate offers an opportunity to freely discuss the pros and cons of the natural gas extraction (hydraulic fracturing), the merits and benefits of the proposed pipelines, and its impact on environment and individual rights. The idea to accept a simplistic approach to a complicated problem belongs to the 19th century mentality and social environments. The time has arrived to face the challenge and explore new opportunities provided to us in the 21st century. And, I'm certain that many bright minds here and in other countries are looking into innovative local or regional energy production so they may not depend on some uncertain energy source from across the ocean thousands of miles away.

During the last few decades, thanks to the Internet and other forms of advanced communication, public awareness significantly increased about the strong link between environmental preservation and public health. We are more aware of environmental and health crisis that face not only America but the entire global community. For the good news, recent innovations in science and renewable energy technologies which are mostly developed by using unlimited fossil fuels (coal, oil and natural gas), allow us to gradually move away from fossil fuels technologies that are detrimental to our health and environment and find new pathways for human happiness, prosperity and liberty. In a free and democratic society, we can no longer be enslaved by established and traditional technologies and practices. We can do much better. This may require some sacrifice but this nation was created by many individual and collective sacrifices and we should have no fear facing the challenge.

4.3 Coal Production: Panacea or Plague?

According to Wikipedia (the free online encyclopedia) Midlothian, the suburban community located west of Richmond was founded over 300 years ago as a coal mining village. It was named for the early 18th-century coal mining enterprises of the Wooldridge brothers, who came from mining villages in East Lothian and West Lothian near Edinburgh, Scotland. This new venture called "Mid-Lothian" produced the first commercially mined coal in the Virginia Colony. Later, the Midlothian-area coal heated the U.S. White House during the presidency of Thomas Jefferson.

History of coal, sometimes called "black diamond", is intertwined with history of human civilization.

Archeological evidence shows coal mining and use in China about 3490 BC. Romans were mining coal in the 2nd century AD. The Aztecs in Americas used coal for fuel in 13th century. However, large-scale coal mining began during the 18th century Industrial Revolution when coal-powered steam engines were built for railways and steamships. Coal became the primary energy source for major industries such as steel production and electricity generation. No doubt, coal use has led to developing modern industries and our today's quality of life could have not happened without coal. At present, over fourth of the world's energy is provided by coal, but there is strong competition from oil, natural gas, nuclear power and other alternative energy sources.

At present, carbon dioxide emission from coal use is cited as a major culprit in global warming and climate change. The major concern for continued coal use is not CO_2 emission but significant environmental degradation due to surface mining in the coalfield counties of Southwest Virginia and elsewhere. Levels of CO_2 emission for coal, petroleum, and natural gas are, respectively 2.117, 1.915, 1.314 pounds/kwh of power generated. Natural gas is promoted as a clean energy source to replace coal. However, in the absence of a national strategy for energy conservation, replacing coal use with large volumes of natural gas will not reduce CO_2 emission to a significant level that could effectively mitigate climate change.

The two basic types of coal mining are deep underground mining and surface mining. Coal seams relatively close to the ground surface (less than approximately 180 ft.) are usually surface mined by strip mining or mountain top removal. Strip mining exposes the coal by removing the overburden (the earth above the coal seam) in long cuts or strips. Soil from subsequent cuts is deposited as fill in the previous cut after coal has been removed. Mountaintop coal mining practice involve removal of mountaintops to expose coal seams, and often disposing of mining overburden (called spoil) in adjacent "valley fills".

Surface mining results in several adverse environmental effects that include change in natural topography, degraded air quality, elimination of natural vegetation, destruction of groundwater aquifers, degradation of streams and rivers and associated living organisms and wildlife habitat. Acid mine drainage, the formation and movement of highly acidic water which can seep into stream/rivers and groundwater aquifers, results in potentially significant impacts on human health and ecosystems. The Surface Mining Control and Reclamation Act of 1977 with the intent to mitigate these effects has been rather unsuccessful.

In early 1980s, as a young researcher, I participated in surface mined land reclamation research projects. However, thirty years later, although these research efforts may have produced numerous graduate theses and journal articles, environmental degradation in coalfield counties continue to persist and the affected communities' quality of life has not improved. The coalfield counties in Southwest Virginia remain economically depressed with significant drinking water and resource management issues. While recognizing coal's tremendous contribution to our modern society, the time has arrived for new thinking to gradually seek better ways and solutions for economic development in far

Southwest Virginia and similar areas where people's livelihood depend on coal mining.

4.4 Strategies for Revitalization of Coalfield Counties in Southwest Virginia

During a 30-year period, I visited all corners of the Commonwealth of Virginia from the mountaintops of coalfield counties where I studied reclamation of abandoned mined lands and drinking water problems, to the coastal Virginia where I studied the feasibility of desalination a study mandated by the Virginia General Assembly. In Virginia, there is no parallel to the natural beauty of the coalfield region with its scenic rivers, mountain ridges, forests and caves. Add to that the region's remarkable history and culture. Unfortunately, during this period, it was painful to witness the economic and social decline of this historic and beautiful region.

The coalfield region in southwest Virginia includes counties of Buchanan, Dickenson, Lee, Russell, Scott, Tazewell and Wise, and the City of Norton. The total area of coalfield counties is 3,225 square miles; the largest, Scott County (538 square miles) and the smallest, Dickenson County (335 square miles). Total population in the seven counties is 201,402 (2012 estimate); highest in Tazewell (44, 289) and lowest in Dickenson (15, 679). The land area of seven counties constitutes about 7.5 % of total Virginia land area and the region's total population corresponds to about 2.4 % of Virginia population. Bold steps including innovative ideas, legislative action and high investment are needed to move the region forward. Here are a few thoughts.

(1) The Virginia Coalfield Economic Development Authority (VCEDA) (Virginia code § 15.2-6000) was created to assist the seven counties and City of Norton to achieve economic stability. It's critical for the VCEDA to streamline the efficiency of governmental organizations and represent the whole region as one entity in order to avoid local competition between counties. The VCEDA eligible use of funds (§ 15.2-6011) are mostly targeted for conventional economic development projects and does not promote innovation. There is a need for the VCEDA to develop stronger public-private partnerships that could invest on innovative ideas for the region.

(2) Focus initial investment on promoting tourism and the hotel industry. For resource generation and supporting small local businesses, aside from urgent and renewed focus on promoting the scenic sites in the region, the area should also be promoted for its history and culture. Plan to build a Coal Museum in the region that introduces the role of coal in the industrial revolution and modern Virginia. As a part of such a museum, preserve a surface mined area and a deep mining area for educational and tourism purposes. See Section 4.3 above 'Coal production: panacea or plague?'. A good reference is also existing coal museums in neighboring states.

(3) In the era of accelerated urbanization, it's critical to preserve and enhance the rural characteristics of the region by incorporating innovative ideas that promote rural economic development. A recent book titled "Rural Future: An Alternative for Society Before 2050 AD" by my mentor

and colleague late Dr. Robert Giles is an excellent reference. In the Introduction, Dr. Giles notes that '*I arrived at the need for a rurally-related corporation, after months of work. Millions of dollars of federal and various matching funds and contributions had done little to help citizens in rural Central Appalachia. Coal mines of international owners moved money as well as coal out of the region, leaving pitiful, small-community "camps" through the region. Mine reclamation addressed some surface areas, and performed some land reshaping. Thousands of acres were un-reclaimed, abandoned. Nutrients have been leached from the soil. flooding is common.*' The book proposes Rural System based on advanced science and designed to provide regional employment, reasonable small community stability, and modern natural resource management, all indexed by profit.

(4) The VCEDA and other public-private partnerships should encourage and support innovative and futuristic green technology educational programs and implementation and creating a Research Park in the region. For example, one possible project would be studying the feasibility of developing solar farms on reclaimed surface mined lands which remain idle. Finally, through legislative action, a highly visible Business-Citizen-Academic Partnership should be formed to develop strategies and attract large-scale investment.

Epilog

Sustainable Living: Paradigm Shift in American Dream

In the 20th century, America achieved a high standard of living. Usually, the standard of living is measured by individual wealth, and the comfort and conveniences of daily life. Several factors contributed to the 20th century American prosperity; work ethics, technical innovations, and abundance of water and energy resources such as oil and coal. However, today we realize that a high standard of living is not necessarily synonymous with a good quality of life. Our current standard of living is based on consumerism and the wastage of natural resources, and much has been sacrificed in terms of quality of life in pursuit of a perceived high standard of living. Furthermore, our 21st century way of living and supporting infrastructures is based on the 19th–20th century social norms and technologies. We are challenged with significant local, regional and global problems such as depletion of natural resources, environmental degradation, deforestation, deterioration of water and transportation infrastructures, health issues, and diminished cultural and recreational resources.

Reevaluating our priorities to move toward protecting and enhancing our quality of life is critically needed.

The concept of "sustainable living" facilitates an opportunity for a paradigm shift in the American Dream that respects quality of life and the formation of a vibrant and livable society. Sustainable living is a holistic way of thinking that connects the dots in all aspects of our daily lives and ensures a safe living environment in harmony with nature for today and for future generations. Sustainable living requires us to change our way of thinking from our current perceptions of what constitutes a high standard of living. Our challenge as a modern society is to shift the gear and turn the wheel of sustainable living in our communities – balancing economic development and societal wellbeing. Sustainable living can be accomplished by initiatives and actions that we can undertake as individuals and as communities.

First, and perhaps the most important action is accepting individual responsibility for our own actions that affect our personal health and local environment in our immediate area. This requires a change in our behavior from wastage to resource conservation wherever possible. Second, we need to promote and implement green infrastructures – water and energy efficient homes and buildings, walkways, bikeways, greenways, parks, and community gardens among others – that support sustainable living and good quality of life in our communities. Third, we should promote and encourage the implementation of decentralized water and energy infrastructures. Decentralized infrastructures are small-scale systems that use locally available water and renewable energy resources to the extent possible. Decentralized infrastructures will preserve natural resources, complement conventional large-scale centralized systems or replace them where appropriate.

Finally, we need to promote widespread educational programs that foster sustainable living at all levels. Policy and decision makers need to become aware of the concept of sustainable living and its positive effect on quality of life and formulate appropriate legislation that removes hurdles to implementation of sustainable living. Local governments need to integrate sustainable living approaches in plans for economic development and infrastructure renovation. Citizen and K-12 educational programs will have significant influence on our way of living. College level curricula should also be updated to include new concepts and designs that promote sustainable living. It should be noted that in our region the Roanoke Valley-Alleghany Regional Commission and the New River Valley Planning District Commission have initiated activities to plan and implement sustainable projects within their jurisdictions. Their initiatives bring together local governments, businesses, academia, nonprofits, and citizen groups to address issues of mutual concern and to promote economic vitality, equity and environmental quality. Their efforts are an excellent example of how the paradigm shift in the American Dream to attain good quality of life can be realized through community-wide decision making and establishing partnerships between diverse elements that drive our wellbeing and economy.

Appendix

Dr. Tamim Younos
Published Books

Alternative Water Sources for Producing Potable Water: Advances in Research and Technology
Editors: Tamim Younos, Juneseok Lee, Tammy E. Parece (2023).
Springer, Handbook of Environmental Chemistry (HEC), Vol. 124, 192 pp.

Synopsis. The central theme of this book is innovations in using alternative water sources for producing potable water. Locally available alternative water sources – rainwater/stormwater, wastewater, seawater/brackish water, atmospheric water – can be effectively used in conjunction with the decentralized and small-scale green water-infrastructure. This volume presents a discussion of alternative water sources capture, system design principles, water treatment processes, integrating renewable energy technologies, economic feasibility, regulation, and challenges facing using alternative water sources. This volume serves as a valuable reference source for water researchers and instruction material for graduate and undergraduate level students in environmental science/engineering, and urban planning/development; and a valuable guide for practice engineers, urban and landscape planners, and water utility personnel involved in holistic and resilient water supply management practices in urban settings and other living environments.

Resilient Water Management Strategies in Urban Settings: Innovations in Decentralized Water Infrastructure Systems
Editors: Tamim Younos, Juneseok Lee, Tammy E. Parece (2022).
Springer Water, 243 pp.

Synopsis. The central theme of this book is innovations in decentralized green water-infrastructure systems (DGWIS). This volume presents a discussion of cross-disciplinary knowledge-base and case studies of DGWIS around the world. Topics include: (1) uses of locally available alternative water sources in urban settings; (2) smart technologies applied to urban water management system; (3) integrating locally available renewable energy use in urban water management system; (4) food-water-energy nexus in urban environments; and (5) decentralized disaster mitigation strategies in urban environments. This volume serves as a reference source for researchers and graduate-level instruction and a valuable guide for practice engineers and landscape planners interested and involved in holistic and resilient water management practices in urban environments.

Socio-Hydrology: The New Paradigm in Resilient Water Management

Editors: Tamim Younos, Tammy E. Parece, Juneseok Lee, Jason Giovannettone, Alaina J. Armel

(2021).

MDPI Publishers, 204 pp.

Synopsis. Socio-hydrology is an emerging cross-disciplinary water science field that integrates the natural and social sciences. This book compiles scientific endeavors and innovations on water research development, education, and applications in the arena of socio-hydrology and broad aspects of water management in the context of socio-hydrology systems from around the globe. Ideas and approaches presented in this book can be significant reference sources for water science programs and an excellent guide for experts involved with the futuristic planning and management of water resources.

Karst Water Environment: Advances in Research, Management and Policy

Editors: Tamim Younos, Madeline Schreiber, Katarina Kosič Ficco (2018).

Springer, Handbook of Environmental Chemistry (HEC), Vol. 68, 274 pp.

Synopsis. Karst aquifers are important sources of drinking water worldwide. This volume presents a discussion of the current state of knowledge on karst science, advances in karst mapping and karst aquifer monitoring technologies, case studies of karst aquifer assessment, and regulatory perspectives on land use and water management in karst environments. It offers valuable reference material for researchers involved in karst science and environmental studies, as well as a guide for experts at governmental agencies, scientists, engineers and other professionals involved in karst aquifer protection and the design of land

and water management systems in karst areas around the globe.

Sustainable Water Management in Urban Environments

Edited by Tamim Younos and Tammy E. Parece (2016). Springer, Handbook of Environmental Chemistry (HEC), Vol. 47, 351 pp.

Synopsis. This 10-chapter book focuses on practical aspects of sustainable water management in urban areas and presents a discussion of key concepts, methodologies, and case studies of innovative and evolving technologies. Topics include: (1) challenges in urban water resiliency; (2) water and energy nexus; (3) integrated urban water management; and (4) water reuse options (black water, gray water, rainwater). This volume serves as a useful reference for students and researchers involved in holistic approaches to water management, and as a valuable guide to experts in governmental agencies as well as planners and engineers concerned with sustainable water management systems in urban environments.

Advances in Watershed Science and Assessment.

Edited by Tamim Younos and Tammy E. Parece (2015).

Springer, Handbook of Environmental Chemistry (HEC), Vol. 33, 292 pp.

Synopsis. This 10-chapter book offers concepts, methods and case studies of innovative and evolving technologies in the area of watershed assessment. Topics discussed include: (1) Development and applications of geospatial, satellite imagery and remote sensing technologies for land monitoring; (2) Development and applications of satellite imagery for monitoring inland water quality; (3) Development and applications of water sensor technologies for real-time monitoring of water quantity and quality; and (4) Advances in biological monitoring and microbial source tracking technologies. This book will be of interest to graduate students and researchers involved in watershed science and environmental studies. Equally, it will serve as a valuable guide to experts in government agencies who are concerned with water-availability and water-quality issues, and engineers and other professionals involved in the design of land- and water-monitoring systems.

Potable Water: Emerging Global Problems and Solutions. Edited by Tamim Younos and Caitlin A. Grady (2014). Springer, Handbook of Environmental Chemistry (HEC), Vol. 30, 233 pp.

Synopsis. This 8-chapter book presents a unique and comprehensive glimpse of current and emerging issues of concern related to potable water. The themes discussed include: (1) historical perspective of the evolution of drinking water science and technology and drinking water standards and regulations; (2) emerging contaminants, water distribution problems and energy demand for water treatment and transportation; and (3) using alternative water sources and methods of water treatment and distribution that could resolve current and emerging global potable problems. This volume will serve as a valuable resource for researchers and environmental engineering students interested in global potable water sustainability and a guide to experts affiliated with international agencies working toward providing safe water to global communities.

Climate Change and Water Resources. Edited by Tamim Younos and Caitlin A. Grady (2013). Springer, Handbook of Environmental Chemistry (HEC), Vol. 25, 231 pp.

Synopsis. This nine-chapter book prepared by international authors highlights various aspects of climate change and water resources. Climate change models and scenarios, particularly those related to precipitation projection, are discussed and uncertainties and data deficiencies that affect the reliability of predictions are identified. The potential impacts of climate change on water resources (including quality) and on crop production are analyzed and adaptation strategies for crop production are offered. Furthermore, case studies of climate change mitigation strategies, such as the reduction of water use and conservation measures in urban environments, are included. This book will serve as a valuable reference work for researchers and students in water and environmental sciences, as well as for governmental agencies and policy makers.

Total Maximum Daily Load: Approaches and Challenges. Edited by Tamim Younos (2005). PennWell Books, Tulsa, Oklahoma. 373 pp.

Synopsis. Total Maximum Daily Load: Approaches and Challenges (9 chapters) presents concepts, approaches, case studies, and applications of the cutting-edge technologies used to develop and implement an effective and innovative TMDL program. Case studies discussed in this book mostly focus on three major causes of water impairment in the United States: bacteria, sediments, and nutrients. The book contains valuable information for anyone involved with pollution control – state and federal water quality agencies, consulting engineering firms, publicly owned treatment works, environmental biologists and chemists, and public health officials.

Advances in Water Monitoring Research. Edited by Tamim Younos (2001)

Water Resources Publications, LLC, 244 pp.

Synopsis. This book presents current issues in water monitoring science, policy, information transfer, research, and advances in technology. Basic concepts discussed in the book include protocols and techniques of ecological restoration and confidence in the restoration process, using sound science in water quality monitoring, and relationship between policy development and monitoring capabilities. Methods of water monitoring and data interpretation for water-quality protection are discussed by presenting several case studies. The book provides an overview of the 21st century water monitoring technologies and their potential for water quality-protection. This book is best suitable as a reference for water monitoring agencies and graduate studies in water monitoring research.